COLLIDER

COLLIDER

The Search for the World's Smallest Particles

Paul Halpern

John Wiley & Sons, Inc.

This book is printed on acid-free paper. ♾

Published by John Wiley & Sons, Inc., Hoboken, New Jersey
Published simultaneously in Canada

Photo credits: Courtesy of the AIP Emilio Segre Visual Archives, 123; courtesy of the AIP Emilio Segre Visual Archives, Physics Today Collection, 91; courtesy of the AIP Emilio Segre Visual Archives, Rutherford Collection, 54; photographs by Maximilien Brice, copyright © CERN, published by permission of CERN, 171, 173; photograph by Joao Pequenao, copyright © CERN, published by permission of CERN, 215; all other photos are from the author's collection.

Library of Congress Cataloging-in-Publication Data:

Halpern, Paul, date.
Collider : the search for world's smallest particles / by Paul Halpern.
p. cm.
Includes bibliographical references and index.
ISBN 978-0-470-28620-3 (cloth)
1. Large Hadron Collider (France and Switzerland) 2. Particles (Nuclear physics)
3. European Organization for Nuclear Research. I. Title.
QC787.P73H35 2009
539.7'376094—dc22

2009007114

Printed in the United States of America

10 9 8 7 6 5 4 3 2 1

To Joseph and Arlene Finston
for their kindness over the years.

Now when chaos had begun to condense, but force and form were not yet manifest, and there was nought named, nought done, who could know its shape?

—*THE KOJIKO* (JAPANESE RECORDS OF ANCIENT MATTERS), TRANSLATED BY BASIL HALL CHAMBERLAIN

Contents

Acknowledgments xi

Prologue: Journey to the Heart of the Large Hadron Collider 1

Introduction: The Machinery of Perfection 7

1 The Secrets of Creation 23

2 The Quest for a Theory of Everything 41

3 Striking Gold: Rutherford's Scattering Experiments 53

4 Smashing Successes: The First Accelerators 75

5 A Compelling Quartet: The Four Fundamental Forces 99

6 A Tale of Two Rings: The Tevatron and the Super Proton Synchrotron 117

7 Deep in the Heart of Texas: The Rise and Fall of the Superconducting Super Collider 147

8 Crashing by Design: Building the Large Hadron Collider 163

9 Denizens of the Dark: Resolving the Mysteries of Dark Matter and Dark Energy 179

10 The Brane Drain: Looking for Portals to Higher Dimensions 191

11 Microscopic Black Holes: A Boon to Science or a Boom for the World? 209

Conclusion: The Future of High-Energy Physics: The International Linear Collider and Beyond 225

Notes 233

Further Reading 241

Index 245

Acknowledgments

Thanks to the many researchers at CERN who took the time to speak with me about their work and show me around the facilities during my visit. In particular, I would like to thank Michael Rijssenbeek, head of the Stony Brook group working on ATLAS, for his hospitality in discussing aspects of the project with me and for organizing a full schedule of meetings that were extremely informative. I would also like to express my deep appreciation to Venetios Polychronakos, Larry Price, Charlie Young, Ashfaq Ahmad, Adam and Katie Yurkewicz, Alexander Khodinov, Jason Farley, Julia Gray, and Jet Goodson.

Many thanks to David Cassidy and Adam Yurkewicz for reading over sections of the text and making suggestions. I would also like to thank J. David Jackson of UC Berkeley and the Lawrence Berkeley National Laboratory for his useful comments. Thanks to David Brin for kindly offering permission to include a quote from his novel *Earth* and to Herman Wouk for graciously allowing me to include a quote from *A Hole in Texas*.

I thank my editors at Wiley, Constance Santisteban and Eric Nelson, for their confidence, help, ideas, and vision, and my agent, Giles Anderson, for his thoughtful guidance and support.

I appreciate the support of my colleagues at the University of the Sciences in Philadelphia for this and other writing projects, including Phil Gerbino, Elia Eschenazi, Jude Kuchinsky, Brian Kirschner, Phyllis Blumberg, Steve Rodrigue, Sergio Freire, Bernard Brunner, Jim Cummings, Ping Cunliffe, Roy Robson, David Traxel, Justin Everett, Deirdre Pettipiece, K. Shwetketu Virbhadra, and many others.

Thanks also to my friends with whom I have exchanged many intriguing ideas throughout the years, including Fred Schuepfer, Michael Erlich, Fran Sugarman, Mitchell and Wendy Kaltz, Simone Zelitch, Doug Buchholz, Bob Jantzen, Robert Clark, Scott Veggeberg, Evan Thomas, Dubravko Klabucar, Elana Lubit, and the late Donald Busky.

Many thanks to my wife, Felicia, and my sons, Aden and Eli, for offering a steady source of love, warmth, and ideas. I also appreciate the support of my parents, Stan and Bunny Halpern; my in-laws, Joseph and Arlene Finston; Shara and Richard Evans; Lane and Jill Hurewitz; Janice, Richard, Jerry, Dolores, Michael, and Maria Antner; Dena and Amritpal Hatton; Aaron Stanbro; Kenneth, Alan, Beth, Tessa, Richard, Anita, Emily, and Jake Halpern; and the rest of my family.

Prologue

Journey to the Heart of the Large Hadron Collider

The ATLAS complex, home to the largest scientific measuring device in the world dedicated to particle physics, offers no hint of its grandeur from street level. From Route de Meyrin, the heavily trafficked road that separates its ground-level structure from CERN's main campus, it looks just like a warehouse near an ordinary filling station. Until I walked through the doors of its main entrance, I wasn't quite sure what to expect.

CERN, the European Organization for Nuclear Research (from the French acronym for Conseil Européen pour la Recherche Nucléaire), located on the Swiss-French border near Geneva, prides itself on its openness. Unlike a military facility, it allows anyone with permission to visit to snap pictures anywhere. Nevertheless, the dangerous and delicate nature of modern particle detectors warrants extremely tight protocols for entering the "caverns" where they are housed.

To tour a site that wasn't yet completely finished, I put on a helmet like that a construction worker would use. My hosts, researchers Larry Price and Charlie Young, were wearing radiation badges—practice for the precautions needed when the beam line would be running. After getting final clearance, they entered the code for the gateway to the inner sanctum, and it electronically opened.

Before journeying underground, we viewed the two enormous shafts used to lower the detector's components more than 350 feet down. I stood on the brink of one of these wells and gazed into the abyss. Awe and vertigo rivaled for my attention as I craned my neck to try to see the bottom.

Above the other well a giant crane was poised to lower parts down below. Transporting the original components of the detector from the various places they were manufactured and then putting them all together within such deep recesses in such a way that their delicate electronics weren't destroyed surely was an incredible undertaking. The meticulous planning for such a complex project was truly phenomenal—and it continues.

We took a speedy elevator ride down to what is called "beam level." Now we were at the same depth as the beam pipes for the Large Hadron Collider (LHC)—the vast ring that will be used to collect protons and other particles, accelerate them in opposite directions, and smash them together at record energies. The ATLAS detector is located at one of the intersection points where protons will collide head on. Half of the detector is above the beam level and half below to accommodate the floods of particles gushing out in all directions from these crashes.

The passages from the elevator to the control rooms and viewing platform are twisty, to provide barriers in case of a radiation leak. Most forms of radiation cannot travel through thick walls. Gauges sample the radiation levels to try to minimize human exposure.

The air in the hallways below seemed a bit stale. Pumped in by means of the ventilation system, it is monitored extremely closely. One of the components of the detector is liquid argon, cooled to only a few degrees above absolute zero—the baseline of temperature. If for some reason the argon heated up suddenly, became gaseous, and leaked, it could rapidly displace all of the breathable air. Warning systems are everywhere; so if such a danger were imminent, workers would be urged to escape via elevator before it was too late.

I finally reached the viewing platform and was astounded by the panorama in front of me. Never before had I seen such a vast array of sparkling metal and electronics, arranged in a long, horizontal cylinder capped with a giant shiny wheel of myriad spokes—nicknamed the "Big Wheel." It was like encountering the largest alien spaceship imaginable—docked in an equally vast spaceport.

If massive new particles are discovered in coming years—such as the Higgs boson, theorized to supply mass to other natural constituents—this could well be the spawning ground. It would come not with the push of a button and a sudden flash but rather through the meticulous statistical analysis of gargantuan quantities of data collected over a long period of time. Not quite as romantic as seeing a new particle suddenly materialize from out of the blue, but statistics, not fireworks, is the way particle discoveries transpire these days.

The inner part of the detector had been hermetically sealed long before my visit, so it was impossible to see inside. Of the outer part, I could barely make out the vast toroidal (doughnut-shaped) magnets that serve to steer the charged elementary particles that escape the inner core. Researchers call these penetrating particles, similar to electrons but more massive, muons. Indeed the main reason for the detector's huge size is to serve as the world's finest muon-catcher. Each "spoke" of the Big Wheel is a muon chamber.

Modern particle detectors such as ATLAS are a bit like successive traps—each designed to catch something different. Particles that slip through one type might be caught in another. The net effect of having an assortment is to capture almost everything that moves.

Suppose a house is full of various vermin and insects. Placing a mousetrap on the kitchen floor might snag some of the rodents but would allow ants to roam free. An ant trap might lure some of those insects but would be inadequate for flies. Similarly, ATLAS consists of multiple layers—each designed to pin down the properties of certain categories of particles.

Some particles, such as electrons and photons (particles of light), are captured by one of the inner layers, called the electromagnetic calorimeter. That's where the light-sensitive liquid argon takes part in measuring their energy released. Other particles, such as protons and neutrons, are stopped by a denser layer just beyond, called the hadronic calorimeter. Those were the inner layers I couldn't see.

For the most part, muons evade the inner detector completely. They are the only charged particles that manage to make it through. That's why the outer envelope of the detector is called the muon system. It consists of what looks like a sideways barrel—centered on the beam line—framed on both sides by two massive "end-caps." These serve to track as many muons as possible—making the detector ideal for any experiments that produce such particles.

As good as ATLAS is at trapping particles, some do manage to make it through all of the hurdles. Extremely lightweight neutral particles, called neutrinos, escape unhindered. There's not much the researchers can do about that, except to calculate the missing energy and momentum. Neutrinos are notoriously hard to detect. Also, because the beam line obviously can't be plugged up, collision products are missed if they head off at low enough angles. Given that most interesting results involve particles whirling off at high angles from the beam line, the lack of information about those traveling close to the beam line is not critical.

A Medusa's head of cables connects the detector's delicate electronics with the outside world, allowing for remote data collection. These connections make it so that relatively few scientists working on the project will need to venture down to the actual detector, once it is operational. Using a system called the Grid, scientists will be able to access and interpret the torrent of information produced via computers located in designated centers around the globe. Then they will look for special correlations, called signatures, corresponding to the Higgs boson and other sought-after particles.

It was humbling to think that the huge artificial cave housing ATLAS comprises but a portion of the LHC's scope. Through one of the cavern walls, beam pipes extend from ATLAS into the formidable tunnel beyond. Miles from where I stood lay other grottos housing various other experiments: CMS, a multipurpose detector with a strong central magnet; ALICE, a specialized detector designed to examine lead-ion collisions; LHCb, another specialized detector focusing on the interactions of what are called bottom quarks; and several other devices.

Back on the surface, I took some time to explore the French countryside above the LHC ring. Most of the seventeen-mile-long circular tunnel lies under a bucolic border region known as Pays de Gex, or "Gex Country." Passport in hand, I caught the bus from Geneva's central station that heads toward the French village of Ferney-Voltaire. According to the map I had with me, the village is roughly situated above one part of the LHC tunnel.

In that quaint locale, where Voltaire once philosophized, mail is still delivered by bicycle. *Boulangeries* bake fresh baguettes according to the ancient tradition, and *fromageries* serve regional cheeses such as tangy Bleu de Gex. Stucco houses, faded yellow or green and capped with burgundy-tiled roofs, line the roads out of town. On the face of it, the community seemed little touched by modernity. The illusion was suddenly broken when I saw a white van turn the corner. Its prominently displayed CERN logo reminded me that this cozy village and its pastoral surroundings play a role in one of the leading scientific endeavors of the twenty-first century.

Back on the Meyrin campus of CERN, I noted a similar juxtaposition of old and new. CERN is a laboratory keenly aware of its history. Its streets are named after a wide range of people who have spent their careers trying to discover the fundamental components of nature—from Democritus to Marie Curie and from James Clerk Maxwell to Albert Einstein. Scattered around its museum area are an assortment of accelerators and detectors of various shapes, sizes, and vintage. Comparing the small early detectors to ATLAS served to highlight the unbelievable progress made in particle physics during the last seventy-five years.

CERN makes good use of many of its historical devices. Particles entering the LHC will first be boosted by several different older accelerators—the earliest built in the 1950s. It is as if the spirits of the past must offer their blessings before the futuristic adventures begin.

With this lesson in mind, before propelling ourselves into modern issues and techniques, we must first boost ourselves up to speed with a look at the history of elementary particles and the methods used to unravel their secrets. I intend this book not just as a guide to the Large Hadron Collider and the extraordinary discoveries likely to be made there, but also as a scientific exploration of humankind's age-old quest to identify nature's fundamental ingredients. Like a ride through a high-energy accelerator, it is a fantastic journey indeed.

Introduction

The Machinery of Perfection

Does not so evident a brotherhood among the atoms point to a common parentage? . . . Does not the entireness of the complex hint at the perfection of the simple? . . . Is it not because originally, and therefore normally, they were One—that now, in all circumstances—at all points—in all directions—by all modes of approach—in all relations and through all conditions—they struggle back to this absolutely, this irrelatively, this unconditionally one?

—EDGAR ALLAN POE *(EUREKA)*

Deep in the human heart is an irrepressible longing for unity. Symmetry and completeness guide our sense of beauty and steer us toward people, places, and things that seem balanced and whole. Architects understand this quest when they make use of geometric principles to design aesthetically pleasing structures. Photographers tap into this yearning when they frame their images to highlight a scene's harmonies. And lovers realize this desire when they seek the comfort of profound connection.

Where might we find perfection? Is it by digging deep into our past to an ancient time before symmetry was shattered? Or is it by digging deep underground, crafting mighty machinery, and shattering particles ourselves—hoping that in the rubble we might somehow find fossils of a paradise lost?

The opposite of beauty is the macabre. Unbalanced things, like broken art or atonal music, make us uneasy. Perhaps no writer better expressed this contrast than Edgar Allan Poe—a master at capturing the gorgeous and the gruesome—who spent much of his final years developing and promoting an attempt to understand the deep unity underlying creation. His prose poem *Eureka* suggests that the universe's original oneness longs to reconstitute itself; like the unsteady House of Usher, it pines for its native soil.

Contemporary physics, a triumph of generations of attempts to map the properties of nature, contains satisfying islands of harmony. Yet it is a discipline with uncomfortable inequities and gaps. Completing the cartography of the cosmos has called for intrepid exploration by today's ablest researchers.

In understanding the forces that steer the universe, and trying to unify them into a single explanation of everything, most of the greatest strides have been made in the past two centuries. In the mid-nineteenth century, the brilliant Scottish physicist James Clerk Maxwell demonstrated that electricity and magnetism are integrally connected, and that the relationship between them could be expressed through a set of four simple equations. Maxwell's equations are so succinct they can literally fit on a T-shirt—as evidenced by a popular fashion choice at many physics conferences. These relationships offer a startling conclusion: all light in the world, from the brilliant yellow hues of sunflowers to the scarlet shades of sunsets, consists of electromagnetic waves—electricity and magnetism working in tandem.

By the early twentieth century, physicists had come to realize that these waves always travel in discrete packets of energy, called photons. These hop at the speed of light between electrically charged objects, causing either attraction or repulsion. Hence, all electromagnetic phenomena in the world, from the turning of a compass needle to the blazing of lightning in the sky, involve the exchange of photons between charged particles.

In addition to electromagnetism, the other known natural interactions include two forces that operate on the nuclear scale—called the weak and strong interactions—and the apple-dropping, planet-guiding force of gravitation. These four forces proscribe all of the ways material objects can attract, repel, or transform. Whenever motion changes—a sudden jolt, a subtle twist, a quiet start, or a screeching halt—it is due to one or more of the four natural interactions.

Each interaction occurs by means of its own exchange particle, or set of particles. An exchange particle works by drawing other particles together, pushing them apart, or changing their properties. It's like a Frisbee game in which the players move closer to catch the Frisbee or step back when it hits them. The process of tossing something back and forth cements the players' connection.

Given Maxwell's fantastic success with marrying electricity and magnetism, many physicists have tried to play matchmaker with the other forces. Like party hosts trying to facilitate connections among their guests, researchers have looked for commonalities as a way of making introductions. Could all four interactions be linked through a mutual set of relationships?

The most successful unification so far has been the melding of electromagnetism and the weak force, performed by American physicists Steven Weinberg and Sheldon Glashow and Pakistani physicist Abdus Salam (each working independently). In unity, these would be known as the electroweak interaction. There were many tricky details that needed to be ironed out, however, before the match could be made.

One of the major issues had to do with a great disparity in the masses of the exchange particles corresponding to each force. While photons have zero mass, the carriers of the weak interaction are bulky—signified by the latter force's much shorter range. The difference between electromagnetic and weak exchanges is a bit like tossing a foam ball across a field and then trying to do the same with a lead bowling ball. The latter would scarcely spend any time in the air before plunging to the ground. With such different volleys, how could the two forces play the same game?

Sometimes, however, inequities emerge from once-balanced situations. Symmetry, as collectors of ancient sculpture know, can be fragile. Perhaps the very early universe, instants after the fiery Big Bang that ushered in its birth, was in a fleeting state of harmony. All forces were perfectly balanced, until some transformation tilted the scales. Equity shattered, and some of the exchange particles became bulkier than others. Could today's unequal forces constitute the culmination of a universal symmetry-breaking process?

In 1964, British physicist Peter Higgs proposed a promising mechanism for spontaneously breaking the universe's initial symmetry. His mechanism requires a special entity, dubbed the Higgs field, that pervades all of space and sets its fundamental energy scale. (A field is a mathematical description of how properties of a force or particle change from point to point.) Within its internal structure is a marker called a phase angle that can point anywhere around a circle. At extremely high temperatures, such as the nascent moments of the universe, the direction of this marker is blurry, like a rapidly spinning roulette wheel. However, as temperatures lower, the roulette wheel freezes, and the marker points to a random direction. The Higgs field's initial symmetry, with all angles being equal, has spontaneously broken to favor a single angle. Because the Higgs field sets the baseline for the vacuum (lowest energy) state of the universe, this transforms during the symmetry breaking from a situation called the true vacuum, in which the lowest energy is zero, to a false vacuum, in which it is nonzero. Following Albert Einstein's famous dictum $E = mc^2$ (energy equals mass times the speed of light squared), the acquired energy becomes mass and is shared among many elementary particles, including the carriers of the weak interaction. In short, the halting of the Higgs field's "roulette wheel" channels mass into the weak exchange (and other) particles and explains why they are bulky while the photons remain massless. With its phenomenal ability to bestow mass on other particles, the Higgs has acquired the nickname the "God particle."

If the Higgs mechanism is correct, a remnant of the original field should exist as a fundamental particle. Because of its high mass—more than a hundred times higher than the protons that compose the cores of hydrogen atoms—it would be seen only during energetic particle events, such as high-energy collisions. Despite decades of searching, this key ingredient for electroweak unification has yet to be found. Hence, the elusive God particle has become the holy grail of contemporary physics.

Aside from the missing Higgs particle, electroweak unification has proven enormously successful. Its importance is such that it is known as the Standard Model. However, much to the physics community's disappointment, attempts to unite the electroweak interactions with the other two forces have yet to bear fruit.

At least the electroweak and strong forces can be written in the same language—the lexicon of quantum mechanics. Developed in the 1920s, quantum mechanics is a powerful toolbox for describing the subatomic realm. Although it offers accurate odds for the outcome of physical events such as scattering (the bouncing of one particle off another) and decay, it frustratingly contains a built-in fuzziness. No matter how hard you try to nail down the precise course of events for natural occurrences on the subatomic scale, you are often left with the flip of a coin or the roll of a die. Einstein detested having to make wagers on what ought to be known with crystal clarity and spent his later years trying to develop an alternative. Nevertheless, like a brilliant but naughty young Mozart, quantum mechanics has offered enough stunning symphonies to hide its lack of decorum.

Physicists who relish exactness tend to turn toward Einstein's own masterpiece, the general theory of relativity, which offers a precise way of describing gravity. Unlike theories of the other interactions, it is deterministic, not probabilistic, and treats space and time as participants rather than just background coordinates. Although there have been numerous attempts, there is no broadly accepted way of placing gravity on a quantum footing. It's like trying to assemble a winning team for a spelling championship but finding that one of the four players, though an expert, speaks a completely different language.

Researchers are left with an odd puzzle. Of the four fundamental interactions, two, electromagnetism and the weak, appear to fit together perfectly. The strong interaction seems as if it ought to fit, but no one has quite been able to match it up. And gravity seems to belong to another set of pieces altogether. How then to reconstruct the original symmetry of the cosmos?

Other areas of asymmetry in contemporary physics include a huge discrepancy between the amount of matter and antimatter (like matter but oppositely charged) in the universe—the former is far more plentiful—and behavioral differences between the particles that build up matter, the fermions, and those that convey forces, the bosons. Like the Montagues and the Capulets, fermions and bosons belong to separate houses with distinct traditions. They like to group themselves in different ways, with fermions demanding more breathing room. Attempts to reconcile the two, in a grand cosmic union called supersymmetry, require that each member of one family have a counterpart in the other. These supersymmetric companions could also help explain a deep conundrum in astronomy: why galaxies appear to be steered by more mass than they visibly contain. Could supersymmetric companions constitute some or all of this dark matter? So far such invisible agents have yet to be found.

Such gaps and discrepancies rattle the human spirit. We like our science to tell a full story, not a shaggy-dog tale. If we can't think of a solid ending, perhaps we haven't imagined hard enough—not that theoretical physicists haven't tried. For each scientific mystery a bevy of possible explanations have sprung up with varying degrees of plausibility.

Recent theoretical efforts to replace elementary particles with vibrating strands or membranes of energy—in what are called string theory and M-theory, respectively—have captured the imagination. These make use of supersymmetry or extra dimensions beyond space and time to explain in an elegant way some of the differences between gravity and the other interactions. An attractive mathematical feature of these theories is that while in prior approaches some calculations involving infinitesimal point particles yielded nonsensical results, using finite strands or membranes removes these problems. Given the difficulties with completing unification through extending the Standard Model, a number of prominent theorists have been won over to the mathematical elegance of these novel approaches. Steven Weinberg, for instance, once remarked, "Strings are the only game in town."[1]

Detractors of string theory and M-theory, on the other hand, question their physical relevance because they contain undetermined values and require unseen dimensions. In the array of all configurations, the real world is just a subset of myriad possibilities. If a theory has enough free parameters, opponents point out, it could represent virtually any kind of particle or interaction. It's like a writer who compares himself to Dickens churning out tens of thousands of pages of uneven prose and instructing an editor to piece together the most Dickensian passages. To echo Truman Capote's famous remark, "that's not writing; that's typing," detractors could well say about string theory, "that's not physics; that's model-making."

Even the most stalwart supporters and fervent detractors would agree that the ultimate test of a theory lies in its experimental verification. So far, for string theory and M-theory, such evidence has been lacking. As noted theorist Bryce DeWitt once told me, "With M-theory I feel dubious about graduate students [going into a field] where there is not one shred of experimental evidence supporting it."[2]

From the 1930s until the mid-1990s, enormous strides were made in particle physics by means of high-energy experimentation with various types of accelerators. An accelerator is a device that uses electric and magnetic fields in tandem to steer particles (such as protons) along tracks or rings while propelling them to higher and higher energies. These particles are then allowed to collide, converting their energy into a multitude of collision products. Following Einstein's equation, the higher the energy at the collision point, the greater the chance of massive particles being produced. While older accelerators used fixed targets, physicists came to realize that smashing particles together head-on produced even higher energies. Accelerators in which particles crash into one another are called colliders.

During those pivotal decades, by using various types of detectors to collect and analyze collision data, researchers were able to identify a zoo of elementary particles. The major theoretical breakthroughs of those times were motivated by a need to organize these particles into families and to understand their decays and interactions. Physicists discovered that all matter particles are either hadrons (subject to the strong force) or leptons (unaffected by the strong force). Protons are examples of hadrons and electrons are examples of leptons. Hadrons are composed of elementary components called quarks—either two or three per particle—bound together by gluons. Quarks come in six varieties called flavors: "down," "up," "strange," "charm," "bottom," and "top." There are also six types of antiquarks, which are similar to quarks but oppositely charged.

In that era of discovery, whenever novel theories were proposed, such as the quark model, researchers set out to test them through further experimentation. Testability lent certain theories a special relevance and clout—allowing them to raise their voices above others and demand to be heard. Upon verification, they could then say, "I told you so."

For example, the top quark, predicted in the 1970s, was identified in 1995 through an analysis of the collision debris of what was then the mightiest accelerator in the world: the Tevatron, at Fermi National Accelerator Laboratory (Fermilab) in Batavia, Illinois. Inaugurated in 1983, the Tevatron propels streams of protons and antiprotons (negatively charged, antimatter counterparts of protons) up to energies of about 1 TeV (Tera electron volts) each before bashing them together. One electron volt is the amount of energy involved in transporting a single electron or proton between the termini of a one-volt battery. Multiply that quantity by one trillion, and the result is 1 TeV—a colossal amount of energy for a minuscule elementary particle.

As it turned out, the discovery of the top quark represented the last major triumph of the Tevatron—and of experimental high-energy physics for some time. Finding the Higgs particle, identifying supersymmetric companions, and other important goals seemed to require more energy than even that mighty machine could muster. Without the opportunity for experimental confirmation the chorus of contending theories began to resemble a meaningless cacophony. Only by building more energetic colliders to test these competing ideas could the malaise of theoretical physics be remedied and significance restored to its voice.

The European Organization for Nuclear Research, better known by the acronym CERN (Conseil Européen pour la Recherche Nucléaire), took up the challenge. With the aim of finding the Higgs particle, discovering possible supersymmetric companions, discerning the nature of dark matter, exploring the possibility of hidden extra dimensions, understanding why there is an abundance of matter over antimatter in the universe, reproducing some of the conditions of the Big Bang, and resolving a host of other critical scientific issues, CERN would channel its resources into the construction of the largest and most powerful accelerator in the world, built at its headquarters near Geneva, Switzerland.

After more than fifteen years of planning and more than eight billion dollars in funding, the Large Hadron Collider (LHC), science's groundbreaking effort to unlock the deepest secrets of particle physics, is finally complete. It is truly the grandest experiment of all time—the pinnacle of humanity's quest for unification. Befitting the pursuit of cosmic grandeur and unity, it is set in a stunning location.

Query a world traveler about locales of striking beauty and harmony, and chances are Switzerland would be high up on the list. From its majestic mountains and crystalline lakes to its quaint cog railways and charming medieval towns, it is hard to imagine a better place to base a search for unification. Indeed the Swiss confederation, uniting inhabitants divided into four different official languages (French, German, Italian, and Romansch), several major religions (Protestant, Catholic, and other faiths), and twenty-six distinct cantons, physically isolated in many cases from one another, represents a model for bringing disparate forces together into a single system. Though in past centuries Switzerland experienced its share of turmoil, in more recent times it has become a haven for peace and neutrality.

As Europe's political frontiers have receded, many scientific roadblocks have fallen as well. The LHC crosses the Swiss-French border with the ease of a diplomat. Its seventeen-mile-long circular underground tunnel, recycled from a retired accelerator called the Large Electron-Positron Collider (LEP), represents a triumph for international cooperation. Only by working in unison, it reminds us, might we discover the secrets of natural unity.

American researchers form a large contingent in the major LHC experiments. They are proud to contribute to such a pivotal venture. Although the United States is not a member of CERN, it donates ample funds toward LHC research. While celebrating Europe's achievements, however, many American physicists still quietly mourn what could have taken place at home.

In 1993, the U.S. Congress voted to cut off funding for what would have been a far bigger, more powerful project, the Superconducting Super Collider (SSC). About fourteen miles of a planned fifty-four-mile tunnel in the region of Waxahachie, Texas, had already been excavated before the plug was pulled. Today that land sits fallow, except for the weed-strewn abandoned buildings on the site. Years of anticipation of novel discoveries were crushed in a single budgetary decision.

President Bill Clinton sent a letter to the House Appropriations Committee expressing his strong concerns: "Abandoning the SSC at this point would signal that the United States is compromising its position of leadership in basic science—a position unquestioned for generations."[3]

Nevertheless, tight purse strings won out over sweeping visions. The cancellation of the SSC shattered the plans of those who had made multiyear commitments to the enterprise and discouraged young researchers from pursuing the field. It would prove a horrendous setback for American high-energy physics, shifting the momentum across the Atlantic.

By delivering a planned 20-TeV burst of energy with each collision, the particle-smasher in Texas would have been energetic enough to conduct a thorough search for the elusive God particle. Perhaps in its hatchery, supersymmetric companion particles would have been born, presenting themselves through their characteristic decay profiles. Dark matter could have made itself known in caverns deep beneath the Texas soil. The ramifications of string theory and other unification models could have been explored. Like the moon landings, these expeditions could have been launched from U.S. soil. With the LHC's completion, the Tevatron will soon be obsolete and no more large American accelerators are planned. What went wrong?

The reason lies with long-term planning and commitment to science, an area where sadly the United States has in recent times often fallen short. Each European member of CERN pledges a certain amount every year, depending on its gross national product. Thus the designers of the LHC could count on designated funding over the many years required to get the enterprise up and running. Already, the upgrades of coming years are being programmed. Foresight and persistence are the keys to the LHC's success.

Not that there haven't been frustrating glitches and delays. Contemporary high-energy physics requires delicate instrumentation that must be aligned perfectly and maintained in extreme environmental conditions such as ultracold temperatures. Despite researchers' best efforts, systems often fail. Originally supposed to go on line in 2005, the LHC wasn't yet ready. Its opening was delayed again in 2007 because of accidental damage to some of its magnets.

On September 10, 2008, proton beams were circulated successfully for the first time around the LHC's large ring. Project leader Lyn Evans and the international team of researchers working at the lab were elated. "It's a fantastic moment," said Evans. "We can now look forward to a new era of understanding about the origins and evolution of the universe."[4]

Nine days later, however, that heady summer of hope screeched to a halt due to a devastating malfunction. Before particle collisions had even been attempted, a faulty electrical connection in the wiring between two magnets heated up, causing the supercooled helium surrounding them to vaporize. Liquid helium is a critical part of the LHC's cooling system that keeps its superconducting magnets functioning properly. In gaseous form, the helium began to leak profusely into the vacuum layer that surrounds the system, thwarting attempts by emergency release valves to channel it off safely. Then came the knockout punch. The flood of helium slammed into the magnets, jostled them out of position, and destroyed more wiring and part of the beam pipe. Upon inspection, technicians realized that it would take many months to repair the damage, recheck the electrical and magnetic systems around the ring, and attempt operations once more. Currently, the LHC is scheduled to go on line in September 2009.

When it is up and running, the LHC will be a marvel to behold—albeit remotely, given that its action will take place well beneath the surface. Burrowed hundreds of feet beneath the earth but only ten feet in diameter, the LHC tunnel will serve as the racetrack for two opposing beams of particles. Steered by more than a thousand gigantic supercooled magnets—the coldest objects on Earth—these particles will race eleven thousand times per second around the loop, traveling up to 99.999999 percent of the speed of light. Reaching energies up to 7 TeV each, the beams will be forced to collide at one of four designated intersection points.

One of these collision sites houses the ATLAS (A Toroidal LHC ApparatuS), detector, a colossal instrument seven stories high (more than half the height of the Statue of Liberty) and spanning 150 feet (half the length of a football field) from end to end. Using sensitive tracking and calorimetry (energy-measuring) devices, it will monitor the debris of protons crashing together in its center, collecting an encyclopedia of data about the by-products of each collision. Halfway around the ring, another general- purpose detector called CMS (Compact Muon Solenoid) will employ alternative tracking and calorimetry systems to similarly collect reams of valuable collision data. At a third site, a specialty detector called LHCb (Large Hadron Collider beauty) will search for the decays of particles containing bottom quarks, with the hope of discovering the reason for the dearth of antimatter in the cosmos. Finally, at a fourth collision site, another specialized detector called ALICE (A Large Ion Collider Experiment) will be reserved for times of the year when lead ions are collided instead of protons. By smashing these together, researchers hope to re-create some of the conditions of the early universe. From each detector, based on careful assessment of the signals for possible new particles, the most promising information will be sent off for analysis via a global computing network called the Grid.

A vast group of researchers from numerous countries around the world will wait eagerly for the LHC results, hoping to find signs of the Higgs, supersymmetric companions, and other long-hoped-for particles. Discovery of any of these would spur a renaissance in physics and an enormous boost for the scientific enterprise—not to mention grounds for a Nobel Prize. The world would celebrate the achievements of those involved in this extraordinary undertaking, including the hardworking Evans and the thousands of workers contributing their vital efforts and ideas to the project.

If the Higgs is found, depending on exactly what mass it turns out to be, the Standard Model could be either confirmed or found in need of major revision. Some supersymmetric alternatives to the Standard Model predict multiple Higgs particles at various energies. Finding evidence of these would be a triumph for supersymmetric theories, especially if other supersymmetric particles are discovered along with them. At the energies of the LHC, most physicists expect to see some new particles. If all goes well, there should be enough for theorists to chew on for many years.

Anticipation is high for the LHC, but so is apprehension. More than any other scientific device in recent memory, there has been an undercurrent of fear that its operation somehow places the Earth or even the whole universe at risk. Disseminated largely through the Internet, these views have caught flame (and been flamed) in numerous blogs and user groups.

The principal culprits for the LHC's supposed threats to our existence include voracious microscopic black holes, ferocious hypothetical particles called "strangelets," magnetic monopoles, and other purported catalysts of doom. Apocalyptic fears are nothing new; many people choose to spend their time worrying about potential calamities such as the collision of asteroids or the evaporation of Earth by a nearby stellar explosion. What is novel about the LHC theories are worries about the world-gobbling powers of theoretical objects that have never been detected and could very well not even exist.

Of these LHC doomsday scenarios, perhaps the most widespread is the notion that the intensity of the collisions would forge a mini–black hole at the collision site that would then grow to Earth-swallowing dimensions by ingesting more and more material, like the gelatinous creature in the horror film *The Blob*. Indeed, there are some theoretical speculations about the creation of microscopic gravitationally intense objects. However, the idea that mini–black holes would act like blobs is an unfortunate misconception; any objects created would be far too minuscule to pose a threat.

Ordinary black holes are the products of the late-stage collapse of heavy stars, at least three times as massive as the Sun. They are called such because of their intense gravitational fields, which are so strong that within an invisible barrier called the event horizon not even light can escape. Typically, the way some black holes accrue matter is if they have companion stars; then they slowly absorb their unfortunate partner's material and grow gradually over time.

Miniature black holes are a hypothetical concept based on the premise of concentrating a large amount of mass in a region the size of an elementary particle. Their event horizons would be so small that the minute bodies would have virtually no gravitational effect on the space a mere fraction of an inch away, let alone elsewhere in the Earth. Moreover, due to a process called Hawking radiation, they would almost immediately evaporate—decaying into other particles. Thus, mini–black holes would scarcely have the opportunity to exist, let alone to grow beyond subatomic proportions. In short, they'd have no chance of destroying even part of the LHC, let alone the Earth.

As Peter Higgs told the *Independent*, "The black hole business has become rather inflated. Even the theorists who are suggesting that mini–black holes are things that could be produced are not predicting black holes large enough to swallow up chunks of the universe. I think the publicity has rather got out of hand and some people have misunderstood it."[5]

In the days before the aborted start-up in September 2008, apocalyptic worries dominated news stories about the collider. "Meet Evans the Atom, Who Will End the World on Wednesday," proclaimed a headline to a piece in the British tabloid *Daily Mail* about the LHC's project leader. The story begins by mentioning that as a child the Welsh-born Evans made explosives with his chemistry set that "blew the fuses of [his] whole house a few times."[6] Could the whole world, the piece considers, be the next thing for him to blow up?

One team of activists, led by former nuclear safety official Walter Wagner, has gone so far as to sue the LHC, pressing for a halt to its operations. In response to public concerns about the LHC's purported dangers, researchers working on the project have issued detailed analyses of potential threats to the planet, demonstrating how none of these are worth fretting about. A 2003 report, "Study of Potentially Dangerous Events during Heavy-ion Collisions at the LHC," found that "classical gravitational effects [of mini–black holes] are completely negligible for LHC energies and luminosities."[7] A follow-up study conducted in 2008 similarly indicated that mini–black holes would present no danger. Both reports pointed out that if such entities could be created, they would have been produced in energetic cosmic rays that continuously rain down upon the Earth. The mere fact that we are here means that anything forged at such energies wouldn't be threatening.

Indeed, the French and Swiss residents living above the particle-smasher seem for the most part calm and happy. CERN prides itself on openness, publishing all of its decisions. It also takes great pains to respect the environment. The land above the collider is largely unspoiled and clean, flourishing with farms and vineyards. If the agency believed there was any reasonable chance that the LHC would imperil the Earth, the device would have been canceled.

Another aspect of the LHC that has caused some consternation as well as excitement is its ability to reproduce some of the conditions believed to have occurred less than one-trillionth of a second after the Big Bang. Does that mean it will actually create a new cosmic explosion and potentially destroy our own universe? Hardly. Rather, it is the energy *per particle* that will resemble conditions from the dawn of time. Forget about astronomical bursts; by human standards the actual energies produced are small—less than a billionth of a dietary calorie per collision! For a subatomic particle, nevertheless, that's one hefty meal. By recording and studying events at such energies, scientists will be able to understand what happened during the actual creation of the cosmos—without risking engendering a new one themselves.

Unraveling the secrets of the origin and composition of the universe is hardly a new venture, although tools such as the LHC make this much easier. Philosophers and scientists have long wondered what happened during the earliest moments of time. What are the smallest things in the world and how do all of the pieces fit together? Could there be a theory of theories explaining all aspects of nature, from the tiniest particles to the cosmos itself? It is wondrous that such longstanding riddles may soon be answered.

1

The Secrets of Creation

When in the height heaven was not named,
And the earth beneath did not yet bear a name,
And the primeval Apsu, who begat them,
And chaos, Tiamut, the mother of them both
Their waters were mingled together,
And no field was formed, no marsh was to be seen;
When of the gods none had been called into being,
And none bore a name, and no destinies were ordained . . .

—ENUMA ELISH, *THE BABYLONIAN EPIC OF CREATION*,
TRANSLATED BY L. W. KING

Hidden among the haze of cosmic dust and radiation, buried in the very soil we walk upon, locked away in the deep structure of everything we see, feel, or touch, lie the secrets of our universal origins. Like the gleaming faces of a beautiful but impenetrable diamond, each facet of creation offers a glimpse of a wonderful, yet inscrutable, unity. With probing intellect, humankind longs to cut through the layers and reach the core of truth that underlies all things. What is the universe made of?

What are the forces that affect our universe? How was the universe created?

Ancient Greek philosophers offered competing explanations of what constitutes the tiniest things. In the fifth century BCE, Leucippus and Democritus, the founders of atomism, argued that materials could be broken down only so far before their basic constituents would be reached. They imagined these smallest, unbreakable pieces, or "atoms," as possessing a variety of shapes and sizes, like an exotic assortment of pebbles and shells.

Another view, proposed by Empedocles, is that everything is a mixture of four elements: fire, water, air, and earth. Aristotle supplemented these with a fifth essence, the void. For two millennia these classical elements were the assumed building blocks of creation until scientific experimentation prodded Europe toward an empirical view of nature.

In his influential book *The Sceptical Chymist*, Robert Boyle (1627–1691) demonstrated that fire, air, earth, and water couldn't realistically be combined to create the extraordinary range of materials on Earth. He argued for a new definition of the term "element" based on the simplest ingredients comprising any substance. Chemists could identify these, he argued, by breaking things down into their most basic parts, rather than through relying on philosophical speculation. Boyle's clever insight challenged experimenters to discover, through a variety of methods, the true chemical elements—familiar to us (in no particular order) as hydrogen, oxygen, carbon, nitrogen, sulphur, and so forth. Whenever children today combine assorted liquids and powders in their chemistry sets, set off bubbling reactions, and concoct colorful, smelly, gooey by-products, they owe a debt to Boyle.

Boyle was an ardent atomist and a meticulous experimenter. Refusing to accept the hypothesis on faith alone, he developed a clever experiment designed to test the concept that materials are made of small particles—which he called corpuscles—with empty space between them. He started with a curved glass tube, exposed to the air on one end and closed on the other. Filling the open end with mercury, he trapped some of the air in the tube and pressed it into a smaller and smaller volume. Then, by slowly removing the mercury, he noted that the trapped air expanded in inverse proportion to its pressure (a relationship now called Boyle's law). He reasoned that this could happen only if the air was made of tiny components separated by gaps.

Manchester chemist John Dalton was an earnest young Quaker whose research about how different substances react with one another and combine led him to the spectacular insight that each chemical element is composed of atoms with distinct characteristics. Dalton was the first, in fact, to use the word "atom" in the modern sense: the smallest component of a chemical element that conveys its properties.

Dalton developed a clever visual shorthand for showing how different atoms combine. He depicted each type of element as a circle with a distinctive mark in the center—for example, hydrogen with a dot, sodium (which he called "soda") with two vertical lines, and silver with the letter "s." Dalton counted twenty elements; today we know of ninety-two natural elements and at least twenty-five more that can be produced artificially. By arranging his circular symbols into various patterns, he showed how compounds such as water and carbon dioxide could be assembled from the "Lego blocks" of elements such as hydrogen, oxygen, and carbon. In what he called the law of multiple proportions, he demonstrated that the elements forming particular substances always combined in the same fixed ratios.

Dalton also attempted to characterize atoms by their relative weights. Although many of his estimates were off, his efforts led to simple arithmetical ways of understanding chemistry. In 1808, Scottish chemist Thomas Thomson combined oxalic acid (a compound of hydrogen, carbon, and oxygen) with several different elements, including strontium and potassium, and produced a variety of salts. Weighing these salts, he found proportionalities corresponding to differences in the elements he used. Thomson's results, published in his book *A System of Chemistry*, helped Dalton's theories gain wide acceptance in the scientific community.

One thing that Dalton's theories couldn't do was predict new elements. Arranging atoms in order of their relative weights didn't offer enough information or impetus for scientists to infer that others existed. It's as if a mother brought three of her sons to a new school to register them and reported only their names and ages. Without saying more about her family, the teachers there would have no reason to believe she had other kids that were older, younger, or in between.

Indeed the family of elements was much larger than Dalton surmised. By the mid-nineteenth century the number of known elements had tripled to about sixty. Curiously, some of these had shared properties—even ones associated with much different atomic weights. For example, sodium and potassium, though separated in terms of weight, seemed to react with other substances in similar ways.

In the late 1860s, Russian chemist Dmitry Mendeleyev decided to write a state-of-the-art chemistry textbook. To illustrate the great progress in atomic theory, he included a chart depicting all of the then-known elements in order of weight. In a bold innovation, he listed the elements in table form with each row representing elements with similar properties. By doing so, he illustrated that elements fall into patterns. Some of the spaces in what became known as the periodic table he left blank, pointing to elements he predicted would later be discovered. He was absolutely right; like a solved Sudoku puzzle, all of the gaps in his table were eventually filled.

Science didn't realize the full significance of Mendeleyev's discovery until the birth of quantum mechanics decades later. The periodic table's patterns reveal that the Democritean term "atom" is really a misnomer; atoms are indeed "breakable." Each atom is a world unto itself governed by laws that supersede Newtonian mechanics. These laws mandate a hierarchy of different kinds of atomic states, akin to the rules of succession for a monarchy. Just as firstborn sons in many kingdoms assume the throne before second-born sons, because of quantum rules, certain types of elements appear in the periodic table before other kinds.

The atom has sometimes been compared to the solar system. While this comparison is simplistic—planetary orbits don't obey quantum rules, for one thing—there are two key commonalities. Both have central objects—the Sun and what is called the atomic nucleus—and both are steered by forces that depend inversely on the squares of distances between objects. An "inverse-square law" means that if the distance between two objects is doubled, their mutual force diminishes by a factor of four; if their distance is tripled, their force weakens ninefold, and so forth. Physicists have found that inverse-square laws are perfect for creating stable systems. Like a well-designed electronic dog collar it allows some wandering away from the house but discourages fleeing the whole property.

While scientists like Boyle, Dalton, and Mendeleyev focused on discovering the ingredients that make up our world, others tried to map out and understand the invisible forces that govern how things interact and transform. Born on Christmas Day in 1642, Sir Isaac Newton possessed an extraordinary gift for finding patterns in nature and discerning the basic rules underlying its dynamics. Newton's laws of mechanics transformed physical science from a cluttered notebook of sundry observations to a methodical masterwork of unprecedented predictive power. They describe how forces—pushes and pulls—affect the journeys through space of all things in creation.

If you describe the positions and velocities of a set of objects and delineate all of the forces acting on them, Newton's laws state unequivocally what would happen to them next. In the absence of force or with forces completely balanced, nonmoving objects would remain at rest and moving objects would continue to move along straight lines at constant speeds—called the state of inertia. If the forces on an object are unbalanced, on the other hand, it would accelerate at a rate proportional to the net force. The extent to which an object accelerates under the influence of a net force defines a physical property called mass. The more massive a body, the harder it is for a given force to change its motion. For example, all other factors being equal, a tow truck's tug would have much less effect on a monstrous eighteen-wheeler than it would on a sleek subcompact car.

Newton famously showed that gravity is a universal force, attracting anything with mass to anything else with mass. The moon, the International Space Station, and a bread crumb pushed off a picnic table by an ornery ant are all attracted to Earth. The more massive the objects, the greater their gravitational attraction. Thus, mass serves two purposes in physics—to characterize the strength of gravity and to determine the accelerating effect of a force. Because mass takes on both roles, it literally cancels out of the equation that determines the effect of gravitational force on acceleration. Therefore bodies accelerate under gravity's influence independent of their masses. If it weren't for the air whooshing by, an aquatic elephant and a mouse up for a challenge would plunge from the high diving board into a swimming pool straight below them at the same rate. The fact that gravitational acceleration doesn't depend on mass places gravity on different footing from any other force in nature.

The concept of attractive forces offers a means by which large objects can build up from smaller ones—at least on astronomical scales. Take scattered bits of slow-moving material, wait long enough for attractive forces to kick in and they'll tend to clump together—assuming they aren't driven apart by even stronger repulsive forces. Attraction offers a natural way for matter to build up from tiny constituents. Therefore it's not surprising that Newton subscribed to the atomist view, believing that all matter, and even light, is made up of minute corpuscles.

In his treatise on optics Newton wrote, "It seems probable to me that God in the beginning formed matter in solid, massy, hard, impenetrable, movable particles of such sizes and figures and with such other properties and in such proportion to space as most conduced to the end for which he formed them. And that these primitive particles being solids, are incomparably harder than any porous bodies compounded of them, even so hard as never to wear or break in pieces; no ordinary power being able to divide what God himself made one in the first creation."[1]

Newton's belief that God fashioned atoms reflected his deeply held religious views about the role of divinity in creation. He believed that an immortal being needed to design, set into motion, and tweak from time to time an otherwise mechanistic universe. His example, in line with the views of the similarly devout Boyle, showed that atomism and religion were compatible.

As Newton demonstrated, the solar system is guided by gravity. Gravity is important on astronomical scales, but it is far too weak a force on small scales to hold atoms together. The force that stabilizes atoms by holding them together is called the electrostatic force, part of what is known as the electromagnetic interaction. While gravity depends on mass, the electrostatic force affects things that have a property called electric charge.

The renowned eighteenth-century American statesman Benjamin Franklin was the first to characterize electric charge as either positive or negative. Influenced by Franklin and Newton, British natural philosopher Joseph Priestley proposed that the electrostatic force, like gravity, obeys an inverse-square law, only depending on charge instead of mass. While gravity always brings objects together, the electrostatic force can be either attractive or repulsive; opposite charges attract and like charges repel. These conjectures were splendidly proven in the 1780s by French physicist Charles-Augustin de Coulomb, for whom the law describing the electrostatic force is named.

Like the electrostatic force, magnetism is another force that can be either attractive or repulsive. The analogue of positive and negative electric charges is north and south magnetic poles. The ancients were familiar with magnetized iron, or lodestone, and knew that by suspending such a material in the air it would naturally align with the north-south direction of Earth. The term "magnetism" derives from the Greek for lodestone, just as "electricity" stems from the Greek for amber, a material that can be easily electrically charged.

Newton's model of forces envisioned them as linking objects by a kind of invisible rope that spans the distance between them. It's like a boy in a first-floor alcove of a church pulling a thin cord that manages to ring a bell in its tower. We call this concept action at a distance. In a way, it is an extension of the Democritean concept of atoms moving in an absolute void. Somehow, two things manage to influence each other without having anything in between to mediate their interaction.

British physicist Michael Faraday found the notion of action at a distance not very intuitive. He proposed the concept of electric and magnetic fields as intermediaries that enable electric and magnetic forces to be conveyed through space. We can think of a field as a kind of ocean that fills all of space. Placing a charge in an electric field or pole in a magnetic field is like an ocean liner disturbing the water around it and disrupting the paths of other boats in its wake. If you were kayaking off the coast of California and suddenly began rocking back and forth, you wouldn't be surprised to see an approaching vessel generating major waves. Similarly, when a charge or pole feels a force it is due to the combined effect on the electric or magnetic field of other charges or poles.

A child playing with a bar magnet in a room illuminated by an electric lightbulb would probably have little inkling that the two phenomena have anything to do with each other. Yet as Danish physicist Hans Christian Ørsted, Faraday, and other nineteenth-century researchers experimentally explored, electrical and magnetic effects can be generated by each other. For example, as Ørsted showed, flipping an electrical switch on and off while placing a compass nearby can deflect its magnetic needle. Conversely, as Faraday demonstrated, jiggling a bar magnet back and forth near a wire can create an electrical current (moving charge) within it—a phenomenon called induction. So a clever enough child could actually light her own play space with her own bar magnet, bulb, and wire.

It took a brilliant physicist, James Clerk Maxwell, to develop the mathematical machinery to unite all electrical and magnetic phenomena in a single theory of electromagnetism. Born in Edinburgh, Scotland, in 1831, Maxwell was raised on a country estate and grew up with a fondness for nature. He loved walking along on the muddy banks of streams and tracing their meandering courses. In his adult life, as a professor at King's College, University of London, he became interested in a different kind of flow, the paths of electric and magnetic field lines fanning out from their sources.

In 1861, convinced that both electricity and magnetism could be explained through the same set of equations, Maxwell synthesized everything that was known at the time about their interconnections. Coulomb's law showed how charge produced an electrostatic force, by way of an electric field. Another law, developed by French physicist André-Marie Ampère based on Ørsted's work, indicated how electric current generated a magnetic field. Faraday's law demonstrated that changing magnetic fields induce electric fields, and another result indicated that changing electric fields create magnetic fields. Maxwell combined these, added a corrective term to Ampère's law, and solved the complete set of equations, resulting in his influential paper "On Physical Lines of Force."

Maxwell's solution demonstrated that whenever charges oscillate, for example, electricity running up and down an antenna, they produce changing electric and magnetic fields propagating through space at right angles to each other. That is, if the electric field strength is changing in the vertical direction, the magnetic field is changing in the horizontal direction, and vice versa. The result is what is called an electromagnetic wave radiating outward from the source like ripples from a stone tossed into a pond.

We can think of electromagnetic radiation as akin to a line dance alternating between men and women, with successive dancers engaged in different hand motions at right angles to each other. Suppose the first dancer is a man who raises his hands up and down. As soon as the woman behind him notices his arms dropping, she shifts her own hands left and right. Then, triggered by her motion, the man behind her lifts his arms up and down, and so forth. In this manner, a wave of alternating hand motions rolls from the front of the line to the back.

Similarly, through successive electric and magnetic "gestures," an electromagnetic wave flows from its source throughout space.

One of the most surprising aspects of Maxwell's discovery was his calculation of the speed of electromagnetic waves. He determined that the theoretical wave velocity matched the speed of light—leading him to the bold conclusion that electromagnetism *is* light. A mystery dating from ancient times was finally resolved—light is not a separate element (the "fire" of classical belief) but rather a radiative effect generated by moving electric charges.

Until the turn of the nineteenth century, science was aware of only optical light: the rainbow of colors that make up the visible spectrum. Each pure color corresponds to a characteristic wavelength and frequency of electromagnetic waves. A wavelength is the distance between two succeeding peaks of the rolling sierra of electromagnetic oscillations. Frequency is the rate per second that peaks of a given wave pass a particular point in space—like someone standing on an express train platform and counting how many carriages zoom by in one second. Because light in the absence of matter always travels at the same speed, as defined by the results of Maxwell's equations, its wavelength and frequency are inversely dependent on each other. The color with the largest wavelength, red, has the lowest frequency—like enormous freight cars taking considerable time to pass a station. Conversely, violet, the color with the shortest wavelength, possesses the highest frequency—akin to a tiny caboose whizzing by.

The visible rainbow comprises but a small segment of the entire electromagnetic spectrum. In 1800, British astronomer William Herschel, best known for discovering the planet Uranus, was measuring the temperatures of various colors and was amazed to find an invisible region beyond the red end of the spectrum that still produced a notable thermometer reading. The low-frequency light he measured just beyond the range of visibility is now called infrared radiation.

The following year, after learning about Herschel's experiment, German physicist Johann Ritter decided to explore the region of the spectrum just beyond violet. He found that invisible rays in that zone, later called ultraviolet radiation, produced a noticeable reaction with the chemical silver chloride, known to react with light.

Radio waves were the next type of electromagnetic radiation to be found. In the late 1880s, inspired by Maxwell's theories, German physicist Heinrich Hertz constructed a dumbbell-shaped transmitter that produced electromagnetic waves of frequencies lower than infrared. A receiver nearby picked up the waves and produced a spark. Measuring the velocity and other properties of the waves, Hertz demonstrated that they were unseen forms of light—thereby confirming Maxwell's hypothesis.

The known spectrum was to expand even further in 1895 with German physicist Wilhelm Roentgen's identification of high-frequency radiation produced by the electrical discharge from a coil enclosed in a glass tube encased in black cardboard. The invisible radiation escaped the tube and case, traveled more than a yard, and induced a chemically coated paper plate to glow. Because of their penetrating ability, X rays, as they came to be called, have proven extremely useful for imaging. They're not the highest frequency light, however. That distinction belongs to gamma rays, identified by French physicist Paul Villard about five years after X rays were discovered and capping off the known electromagnetic spectrum.

The picture of light described by Maxwell's equations bears little resemblance to the Newtonian idea of corpuscles. Rather, it links electromagnetic radiation with other wave phenomena such as seismic vibrations, ocean waves, and sound—each involving oscillations in a material medium. This raises the natural question, What is the medium for light? Could light travel through absolute vacuum?

Many nineteenth-century physicists believed in a dilute substance, called ether, filling all of space and serving as the conduit for luminous vibrations. One prediction of that hypothesis is that light's measured speed should vary with the direction of the ether wind. A famous 1887 experiment by American researchers

Albert Michelson and Edward Morley disproved the ether hypothesis by showing that the speed of light is the same in all directions. Still, given the compelling analogy to material waves, it was hard for the scientific community to accept that light is able to move through sheer emptiness.

The constancy of the speed of light in a vacuum raised another critical question. In a scenario pondered by the young Albert Einstein, what would happen if someone managed to chase and catch up with a light wave? Would it appear static, like a deer frozen in a car's headlights? In other words, in that case would the measured speed of light be zero? That's what Newtonian mechanics predicts, because if two things are at the same speed, they should seem to each other not to be moving. However, Maxwell's equations make no provision for the velocity of the observer. The speed of light always flashes at the same value, lit up by the indelible connections between electric and magnetic fluctuations. Einstein would devote much of his youthful creativity to resolving this seeming contradiction.

Einstein's special theory of relativity, published in 1905, cleared up this mystery. He modified Newtonian mechanics through extra factors that stretch out time intervals and shrink spatial distances for travelers moving close to light speed. These two factors—known respectively as time dilation and length contraction—balance in a way that renders the measured speed of light the same for all observers. Strangely, they make the passage of time and the measurement of length dependent on how fast an observer happens to be moving, but that's the price Einstein realized he had to pay to reconcile Maxwell's equations with the physics of motion.

Einstein found that in redefining distance, time, and velocity, he also had to rework other properties from Newtonian physics. For example, he broadened the concept of mass to encompass relativistic mass as well as rest mass. While rest mass is the inherent amount of matter an object possesses, changing only if material is added or subtracted, relativistic mass depends on the object's velocity. An initially nonmoving chunk of matter starts out with its rest mass and acquires a greater and greater relativistic mass if it speeds up faster and faster. Einstein determined that he could equate the total energy of an object with its relativistic mass times the speed of light squared. This famous formula, $E = mc^2$, implied that under the right circumstances mass and energy could transform into each other, like ice into water.

Yet another question to which Einstein would apply his legendary intellect concerned whether light's energy depends solely on its brightness or has something to do with its frequency. The traditional theory of waves associates their energy with their amount of vibration; waves rising higher carry more energy than flatter waves. For example, pounding on a drum harder produces stronger vibrations that result in a louder, more energetic sound. Just as loudness represents the intensity of sound, a function of the amplitude or height of its waves, brightness characterizes the intensity of light, similarly related to its wave amplitude.

An object that absorbs light perfectly is called a blackbody. Heat up a blackbody box (a carton covered with dark paper, say) and like any hot object it starts to radiate. If you assume that this radiation is in the form of electromagnetic waves distributed over every possible frequency and attempt to figure out how much of each frequency is actually produced, a problem arises. Just as more folded napkins can fit into a carton than unfolded napkins, more types of low-wavelength vibrations can fit into a box than high-wavelength vibrations. Hence, calculations based on classical wave models predict that armies of low-wavelength modes would seize the bulk of the available energy compared to the paltry set of high-wavelength vibrations. Thus, the radiation from the box would be skewed toward low-wavelength high-frequency waves such as ultraviolet and beyond. This prediction, called the ultraviolet catastrophe, is not what really happens, of course; otherwise if you heat up a food container that happens to have a dark coating and set it on a kitchen table, it would start emitting UV radiation like a tanning bed, harmful X rays, and even lethal gamma rays. Clearly the presumption that light is precisely like a classical wave is a recipe for disaster!

In 1900, German physicist Max Planck developed a mathematical solution to the blackbody mystery. In contrast to the classical wave picture of light, which imagines it delivering energy proportional to its brightness, he proposed that light energy comes in discrete packages, called "quanta" (plural of "quantum," the Greek for package), with the amount of energy proportional to the light's frequency. The constant of proportionality is now called Planck's constant. Planck's proposal eliminated the ultraviolet catastrophe by channeling energy into lower frequencies.

Five years later, Einstein incorporated the quantum idea into a remarkable solution of a phenomenon called the photoelectric effect. The photoelectric effect involves what happens when light shines on a metal, releasing electrons (negatively charged particles) in the process. Einstein showed that the light delivers energy to the electrons in discrete quanta. In other words, light has particlelike as well as wavelike qualities. His solution offered the fledgling steps toward a full quantum theory of matter and energy. With his special relativity, energy-matter equivalence, and photoelectric papers all published in 1905, no wonder it is known as Einstein's "miracle year."

Soon thereafter, Russian German mathematician Hermann Minkowski recast special relativity in an extraordinary fashion. By labeling time the fourth dimension—supplementing the spatial dimensions of length, width, and height—he noticed that Einstein's theory took on a much simpler form. Abolishing space and time as separate entities, Minkowski declared the birth of four-dimensional "space-time."

Einstein soon realized that space-time would be a fine canvas for sketching a new theory of gravity. Though he recognized the success of Newton's theory, Einstein wished to construct a purely local explanation based on the geometry of space-time itself. He made use of the independence of gravitational acceleration on mass to formulate what he called the equivalence principle: a statement that there is no physical distinction between free-falling objects and those at rest. From this insight he found a way to match up the local effect of gravity in every region of space-time with the geometry of that region. Matter, he proposed, bends space-time geometry. This warping forces objects in the vicinity to follow curved paths. For example, due to the Sun's distortion of space-time, the Earth must travel in an elliptical orbit around it. Thus, space-time curvature, rather than invisible, distant pulls, is the origin of gravity. Einstein published his masterful gravitational description—called the general theory of relativity—in 1915.

A basic analogy illustrates the general relativistic connection between material and form. Consider space-time to be like a mattress. If nothing is resting on it, it is perfectly flat. Now suppose a sleepy elephant decides to take a nap. When it lies down, the mattress would sag. Any peanuts the elephant might have dropped on its surface while snacking would travel in curved paths due to its distortion. Similarly, because the Sun presses down on the solar system's space-time "mattress," all of the planets in the Sun's vicinity must journey along curved orbits around it.

One of the outstanding features of general relativity is that it offers clues as to the origin of the universe. Coupled with astronomical evidence, it shows that there was a beginning of time when the cosmos was extremely hot and dense. Over billions of years, space expanded from minute proportions to scales large enough to accommodate more than one hundred billion galaxies, each containing billions to hundreds of billions of stars.

The idea of spatial expansion surprised Einstein, who expected that his theory of gravity would be consistent with a static universe. Inserting a sample distribution of matter into the equations of general relativity, he was astonished to find the resulting geometry to be unstable—expanding or contracting with just the tiniest nudge. It was like a rickety building that would topple over with the mere hint of a breeze. Given his expectations for large-scale constancy, that wouldn't do. To stabilize his theory he added an extra term, called the cosmological constant, whose purpose was essentially to serve as a kind of "antigravity"—preventing things on the largest scale from clumping together too much.

Then in 1929, American astronomer Edwin Hubble made an astonishing discovery. Data taken at the Mount Wilson Observatory in Southern California demonstrated that all of the other galaxies in the universe, except for the relatively close ones, are moving away from our own Milky Way galaxy. This showed that space is expanding. Extrapolating backward in time led many researchers to conclude that the universe was once far, far smaller than it is today—a proposal later dubbed the Big Bang theory.

Once he realized the implications of Hubble's findings, Einstein discarded the cosmological constant term, calling it his "greatest blunder." The result was a theory that modeled a steadily growing universe. As Russian theorist Alexander Friedman had demonstrated in previous work, depending on the density of the universe compared to a critical value, this growth would either continue forever or reverse course someday. Recent astronomical results have indicated, however, that not only is the universe's expansion continuing, it is actually speeding up. Consequently, some theorists have suggested a revival of the cosmological constant as a possible explanation of universal acceleration.

Today, thanks to detailed measurement of the background radiation left over from the Big Bang, the scientific community understands many aspects of how the early universe developed and acquired structure. This radiation was released when atoms first formed and subsequently cooled as space expanded. Hence, it offers a snapshot of the infant universe, showing which regions were denser and which were sparser. Einstein's theoretical achievements combined with modern astronomical observations have opened a window into the past—enabling scientists to speak with authority about what happened just seconds after the dawn of time.

Science has made incredible strides in answering many of the fundamental questions about the cosmos. Our sophisticated understanding of the building blocks of matter, the fundamental forces, and the origins of the universe reflect astonishing progress in chemistry, physics, astronomy, and related fields. Yet our curiosity compels us to press even further—to attempt to roll back the hands of time to the nascent instants of creation, a mere trillionth of a second after the Big Bang, and understand the fundamental principles underlying all things.

Since we cannot revisit the Big Bang, the Large Hadron Collider (LHC) will serve as a way of reproducing some of its fiery conditions through high-energy particle collisions. Through the relativistic transformation of energy into mass, it will offer the possibility of spawning particles that existed during the embryonic moments of physical reality. It will also offer the prospect of exploring common origins of the natural forces. Thus, from the chaotic aftermath of particles smashing together at near light speeds, we could possibly unlock the secrets of a lost unity.

Distilling novel ideas from turbulence is nothing new to the people of Geneva. Only six miles southeast of the LHC is Geneva's stunningly beautiful old town. The historic streets and squares, where Jean Calvin once preached religious independence and Jean-Jacques Rousseau once taught about social contracts, are used to all manner of revolutionary currents. Soon Geneva could witness yet another revolution, this time in humanity's comprehension of the fundamental nature of the cosmos.

2

The Quest for a Theory of Everything

What immortal hand or eye
Dare frame thy fearful symmetry?

—WILLIAM BLAKE ("THE TYGER," 1794)

In the heart of Geneva's old town stands the majestic Cathedral of St. Pierre. Between 1160 and 1232, it was constructed in the austere, measured Romanesque style characteristic of the times. Emphasizing the basic unity of God's plan, its vaulting arches and lofty towers were planned to form a careful equilibrium—the left side balancing the right.

Over the ages, the shifting currents of religious belief eroded the cathedral's original design. In the sixteenth century, the Reformation ushered in a fanatical desecration of its interior artwork, including the destruction of statues and the whitewashing of frescoes painted on the walls. Adding to the architectural jumble, the original frontage was replaced with a neoclassical facade in 1750.

Many physicists believe that the universe was once a simple cathedral, elegant and balanced. According to this view, like a perfectly fashioned nave, the cosmos began with an equal mixture of opposites—positive and negative charge, matter and antimatter, leptons and quarks, fermions and bosons, and so forth. As the universe cooled down from its initial hot, ultracompact state, these symmetries spontaneously broke down. Presently the cosmos is thereby a bit of a jumble, like St. Pierre's.

One of the principal missions of the Large Hadron Collider (LHC) involves a kind of archeological expedition—attempting to piece together some of nature's original symmetries. Searching for these symmetries pertains to the ultimate quest to unify all of the particles and forces in the universe under a single umbrella. Looking back to the first moments of the universe could provide the answer. The LHC's extraordinary energies, when applied on the particle scale, reproduce some of the conditions a tiny fraction of a second after the Big Bang. On a minuscule level, it offers a kind of journey back in time.

The LHC isn't re-creating the actual Big Bang itself. Although the LHC's experiments involve comparable energy *per particle*, they produce incomparably less energy overall. It's like pouring a thimbleful of water on a smidgen of sand to test beach erosion; although it might simulate the effect of a lap of the ocean on a bit of the beach, it would hardly reproduce the might of the entire Pacific.

Theoretical clues as to the original state of symmetry present themselves through conserved or near-conserved quantities in nature today. This near-symmetry led to the development of the elegant Standard Model. The Standard Model is a mathematical way of expressing two of nature's fundamental forces—electromagnetism and the weak interaction (a cause of certain types of radiation). There have been numerous attempts to unify these interactions with either or both of the two other natural forces, the strong interaction (that binds nuclei together) and gravity.

For example, Einstein spent the final decades of his life trying to unite electromagnetism with gravity through various extensions of general relativity. He believed that the laws of nature offered subtle signs of an original harmony. These hidden universal principles, he hoped, would eventually reveal themselves through diligent mathematical exploration. Alas, all of his efforts were to no avail. He died in 1955 without finding a satisfactory resolution of his quest.

In the decades after Einstein's death, the Standard Model of the electroweak interaction—as the merger of electromagnetism with the weak interaction is called—took shape as the only fully successful unification model to date. Even this matchup took much hard work and creative thinking. Under mundane conditions, electromagnetism and the weak interaction have several noticeable distinctions. Electromagnetism acts over an incredibly wide range of distances, from the minute scale of atoms to the colossal spans of lightning bolts. The weak interaction, in comparison, acts exclusively on the subatomic level. Moreover, while the electromagnetic force can bring together or push apart charged particles, it never alters their actual charges or identities. Thus, a positively charged proton tugging on a negatively charged electron each remain just that. In contrast, the weak interaction, in its typical dealings, acts like a minuscule marauder, robbing particles of their charge and other properties. For example, it causes beta decay, a process that involves the transformation of a neutral neutron into a proton (along with other particles).

Clever theorists noted, however, that neutrons and protons have similar (but not identical) mass. They pondered, therefore, if the transformation of one into the other could be a one-time symmetry that somehow cracked. Like the Liberty Bell, perhaps it once rang like others forged in the same foundry but then acquired imperfections over the course of time. Could it be that electromagnetism and the weak interaction were born twins but had distinct formative experiences?

The concept of spontaneous symmetry breaking, on which the Standard Model is based, stems from a completely different field of physics: the study of superconductivity. Certain materials, when extremely cold, lose all resistance and conduct electricity perfectly. The "supercurrents" block external magnetic fields from entering and keep internal fields intact. Superconducting magnets are used throughout the LHC to generate the ultra-high fields needed to steer particles around the ring and to keep them focused in tight bunches.

In 1957, John Bardeen, Leon Cooper, and J. Robert Schrieffer (BCS) developed a successful quantum theory of how materials organize themselves to produce such a superconducting state. The theory relies on special correlations between electrons, known as Cooper pairs. The paired electrons organize themselves and, like dutiful soldiers, march in unison. Hence, they are able to overcome all resistance and become perfect electrical conductors.

The reason *paired* electrons are able to move in lockstep while *single* electrons cannot has to do with a feature called the Pauli exclusion principle. Elementary particles fall into two different categories called fermions and bosons. Electrons (if not in Cooper pairs) are an example of fermions and photons are an example of bosons. The Pauli exclusion principle, a critical rule of quantum mechanics, states that no two fermions can share the same quantum state. The principle doesn't apply to bosons, for which any quantity can occupy the same state. It's like a summer camp for which any number of kids (the bosons) are allowed to share the same bunk, but counselors (the fermions) each get a room to themselves. Naturally the former would be much more clustered than the latter—explaining why bosons can act in tandem more easily. Although composed of two fermions each, Cooper pairs behave like bosons, explaining their lockstep behavior.

The mortal enemy of superconductivity is heat. At a sufficiently high temperature, depending on the material, synchronized motion breaks down and superconductivity reverts to normal electrical behavior. The changeover resembles the transformation from crystalline ice into liquid water and is called a phase transition.

Four years after the BCS theory was published, Japanese-born physicist Yoichiro Nambu cleverly demonstrated that its assumptions could similarly describe how symmetry could spontaneously break down in particle physics. With a lowering of temperature, such as in the instants after the Big Bang, a phase transition could occur in which bosons suddenly synchronize themselves and turn aimless behavior into coordinated patterns. He would share the 2008 Nobel Prize for this key finding.

Then, in 1964, British physicist Peter Higgs proposed a new type of boson that would acquire mass through a special kind of spontaneous symmetry breaking. In acquiring mass it would also bestow mass on other particles. Although the boson ended up being named after him, there were several other similar mechanisms proposed independently at the time, including in one paper by Gerald Guralnik, C. Richard Hagen, and Tom Kibble, and another by François Englert and Robert Brout.

In quantum physics, energetic fields take shape due to the potentials they live in. A potential is a kind of a slope, well, or barrier that delineates how energy changes with position. A clifflike potential, for instance, represents a much steeper energy transformation than a plateaulike potential. Higgs assigned his boson a peculiar potential shaped like the bottom of a basin for higher temperatures but like the rim of a basin for lower temperatures. By lowering the temperature below a critical value, the boson is forced from the center of the basin (a zero-energy state, known as the true vacuum), to a place along the rim (a non–zero energy state, called the false vacuum). The arbitrary place on the rim where the Higgs boson ends up—indicating its phase (a type of internal parameter that can assume different angles, like the hands on a clock)—locks in the phase of the ground state of all of space. That's because, unlike an individual particle, a vacuum must be unique and can't have different phases at each point. Hence, the original symmetry is spontaneously broken.

To envision this situation, consider a new property development that is a checkerboard of perfectly square tracts. Before houses are built on the land, each tract is absolutely symmetric with no features distinguishing the north side from the south. Now suppose that a regional ordinance mandates that houses must be spaced a certain distance apart. If the house to be built sits precisely in the center of one of the tracts, then all of the others could follow, and the tracts would each remain symmetric. That's similar to the high-temperature case for Higgs bosons. However, suppose the first house appears in the southwest corner of one of the tracts. The neighboring houses, required to be a specific distance from others, would have to do the same. Eventually, all of the tracts would be occupied with houses on their southwest corners—breaking their original symmetry because of a single, arbitrary, local decision. If the first house had been built in the northeast corner instead, perhaps that would have set the overall trend as well. Similarly, the phase choice of a Higgs boson locally sets the overall phase globally.

As Higgs demonstrated, once the boson field's phase is set, it acquires a mass associated with its nonzero energy. This mass does not arise out of the blue; rather it represents the transformation of energy into mass described by Einstein's special theory of relativity that takes place during the transition between the different vacuum states. Moreover, the Higgs boson interacts with other particles and bestows them with their own masses. Thus, the Higgs could well have set the masses for all of the massive particles in the universe. Because of its ability to seemingly pull mass out of the blue, it has been nicknamed the "God particle"—an epithet with which Higgs himself is not particularly comfortable. It took awhile for the modest professor to get used to a particle named after himself, let alone one assigned divine features.

Higgs's idea was so radical that his original paper was rejected by the leading European journal, *Physics Letters*. He later recalled his disappointment:

> I was indignant. I believed that what I had shown could have important consequences in particle physics. Later my colleague . . . at CERN told me that the theorists there did not see the point of what I had done . . .
>
> Realizing that my paper had been short on salestalk, I rewrote it with the addition of two extra paragraphs, one

of which discussed spontaneous breaking of the currently fashionable SU(3) flavor [quark type] symmetry, and sent it to *Physical Review Letters*. . . . This time it was accepted.[1]

Higgs's paper stimulated a novel look at unifying the electromagnetic and weak forces into a single theory. The critical idea is that these forces are conveyed by means of a quartet of exchange particles, three of which acquire mass through the Higgs mechanism. An exchange particle is a boson that cements the connection between a set of matter particles, causing attraction, repulsion, or transformation. The more massive the exchange particle, the shorter the range of the corresponding interaction.

As the carriers of a force with indefinite range—electromagnetism—photons are massless. Moreover, because they don't affect the charge of interacting particles, they are electrically neutral. Two of the exchange particles for the weak force, however, called the W^+ and W^- bosons, are charged and massive, reflecting the properties of the interaction they convey—charge-transforming and short-ranged. There is also a neutral weak force carrier, called the Z boson. Its existence was proposed in 1960 by Harvard theorist Sheldon Glashow. All three of the weak exchange particles have since been found.

Once Higgs's mechanism was included, along with representations of the exchange particles and fields representing various types of matter, all of the pieces for uniting electromagnetism with the weak interaction snapped into place. In 1967, American physicist Steven Weinberg, working at MIT, and Pakistani physicist Abdus Salam, working at Cambridge, independently developed a successful theory of electroweak unification. It is a masterful theory—the crowning achievement of decades of experimental and theoretical explorations of the nature of subatomic particles. Its designation as the Standard Model is a recognition of its extraordinary importance.

According to theoretical predictions, a remnant of the original Higgs field ought to be leftover and detectable. Surprisingly, despite several decades of experimental investigations at that energy, the Higgs boson has yet to be found. Through the LHC, the physics community hopes at long last to identify the Higgs boson and establish the Standard Model on debt-free grounds.

LHC researchers are fully aware that the Standard Model could prove to be incomplete. Too many mysteries remain about inequities in the universe for the Standard Model to be the be-all and end-all, anyway. Because the Higgs has yet to be found and other interactions are yet to be united satisfactorily, among other things, many physicists today are agnostic about the Standard Model's ultimate validity. Although it has been enormously successful in explaining most particle phenomena, like many beautifully painted old frescoes it has acquired cracks.

At the LHC, researchers often consider the Standard Model predictions along with several alternatives, hoping that experimental results will distinguish among the possibilities. For example, experimenters are preparing themselves for Higgs bosons of higher mass than the Standard Model forecasts or even, as some theories foretell, a triplet of Higgs particles. As midwives to a possible impending birth, they need to ready themselves for a variety of natal scenarios.

Of the unification models that have emerged in the past few decades, the most popular by far has been string theory. String theory envisions the most elementary constituents of nature as incredibly minuscule (less than 10^{-33} inches, called the Planck length) vibrating strands of energy instead of point particles (as envisioned in the Standard Model). Thus they have a finite, but unobservably small, rather than infinitesimal extent. An immediate mathematical advantage is that any equations that include inverse length are finite, rather than infinite. This helps avoid certain mathematical maladies that plague standard quantum field theory, in which particular terms become indefinitely and unrealistically large.

String theory is sometimes called the Theory of Everything (TOE) because it purports to include all known interactions. Its finiteness makes it particularly adept to handle gravitation, which has resisted all previous attempts at inclusion within a unified theory—including Einstein's famous effort. Critics, however, have pointed to string theory's embarrassment of riches. Not only could it potentially include the Standard Model as one of its subsets, but it also seems to encompass myriad physically unrealistic configurations. Therefore one of the long-term goals of string theory is narrowing it down to a single TOE that precisely models our own universe.

According to string theory, different fields and particles are distinct modes of energetic vibrations. If a guitar is out of tune, you can try to tighten its strings. Similarly the energetic vibrations of string theory respond to changes in tension. They also exhibit harmonic patterns, like the overtones that enrich a musical composition. These string configurations correspond to the assorted masses, spins, and other properties of various types of constituents.

String theory started out as solely a model of the strong interaction. In that guise, it encompassed only carriers of force—in other words, bosons. Bosonic string theory could never describe material particles, represented at the tiniest level by fermions. Theorists were motivated to find a way of describing fermions too and model the stuff of matter along with the agents of attraction.

To include fermionic strings along with bosonic strings, physicist Pierre Ramond of the University of Florida proposed the concept of supersymmetry in 1971. Ramond's notion of a transformation connecting matter with force quickly caught fire and ignited the interest of all manner of theorists—even those unenthusiastic about strings themselves. A symmetry uniting fermions with bosons seemed to be the ultimate particle democracy.

Moreover, unlike conventional quantum field theory, like the Standard Model, supersymmetry makes ample room for gravity. For the first time in the history of quantum physics, gravity seemed within reach of incorporation into a unified field theory. Suddenly, Einstein's dormant quest for unification sprang back to life like an antique car equipped with a roaring new engine.

Propelled by the dynamo of supersymmetry, which acquired the nickname "SUSY," field theorists who believed in its power found themselves with the choice of several different routes.

One was to press forward with superstrings, the supersymmetric theory of strings, and to explore their fundamental properties, hoping these would match up with observed aspects of elementary particles. In 1984, an important result by Michael Green and John Schwarz showing that superstring theory lacks certain mathematical blemishes called "anomalies" was cause for celebration. Superstring theory's vibrant vehicle seemed to gleam even more.

One challenge for those taking the fundamental route, however, was making their case to experimentalists. String theory calculations are often tricky and involve many free parameters. These could be adjusted to accommodate a wide range of predictions. Also, until Ed Witten and other theorists showed their equivalence in the mid-1990s, string theories appeared to come in several different varieties. Given such a multiplicity of parameters and theories, researchers were unsure exactly what to test. At any rate, a realm so tiny that nuclei loomed like galaxies in comparison seemed virtually impossible to explore.

Moreover, superstrings are mathematically consistent only if they live in a world of ten dimensions or more. To accommodate the fact that people observe only three dimensions of space and the dimension of time, theorists recalled an idea developed by Swedish physicist Oskar Klein in the 1920s and proposed that six of the dimensions are curled into a ball so tiny that it cannot physically be observed. That worked well mathematically but offered no incentive for experimenters to try to probe the theory. Given the inability to obtain experimental proof, string theory's skeptics—Glashow and Richard Feynman, among the prominent examples—argued that it remained on shaky ground.

Laboratory researchers were enticed to a greater extent by a more conservative application of supersymmetry, called the Minimal Supersymmetric Standard Model (MSSM). Proposed in 1981 by Stanford University physicist Savas Dimopoulos, along with Howard Georgi, it offered a way of extending the Standard Model to include additional fields with the goal of preparing it to be part of a greater unified theory. These fields included supersymmetric companion particles, the lightest of which could potentially be seen in the lab.

The ultimate unification would include gravity. Yet gravity is far weaker than the other forces. Tracing back the history of the universe to a time when gravity could have been comparable in strength to its kindred interactions requires us to ponder its conditions less than 10^{-43} seconds after the Big Bang. At that moment, called the Planck time, the cosmos would have been unimaginably hot and compact, so much so that quantum mechanical principles pertaining to nature's smallest scales would apply to the realm of gravitation. For the briefest instant, the disparate worlds of general relativity and quantum mechanics would be joined through the shotgun marriage of quantum gravity.

Because unification of all of the natural forces would have taken place at such high energies, the particles involved would be extremely heavy. Their mass would be a quadrillion times what could possibly be found at the LHC. Interacting with the Higgs, the Planck scale particles would tug its energy so high as to destabilize the Standard Model. In particular, it would render the weak interaction in theory much feebler than actually observed.

To avoid such a catastrophe, Dimopoulos and Georgi made use of auspicious mathematical cancellations that occurred when they constructed a supersymmetric description of a unified field theory. The cancellations negated the influence of higher mass terms and protected the Higgs from being yanked to unrealistic energies. One caveat is that the Higgs itself would be replaced by a family of such particles—charged along with neutral—including a supersymmetric companion called the higgsino.

If some of the low-mass supersymmetric companions are found, they would offer vital clues as to what lies beyond the Standard Model. They would reveal whether the MSSM or other extensions are correct, and if so, help tune the values of their unspecified parameters (the MSSM has more than a hundred). Ultimately, the findings could provide a valuable hint as to what string theory (or another unified field theory) might look like at much higher energies.

Because string theory has so many different possible configurations and its full energy could be well beyond reach, it is unlikely, however, that any LHC results would either confirm or disprove string theory altogether. At best, they would simply offer more information about string theory's limits and constraints. The experimental discovery of supersymmetry, for instance, would not validate string theory but might assure some of its proponents that they are on the right track.

One of the groups most desperately seeking SUSY consists of researchers trying to resolve one of the deepest dilemmas facing science today: the missing matter mystery. Astronomers are puzzled by unseen matter, scattered throughout the universe, that makes its presence known only through gravitational tugs—for example, through extra forces on stars in the outer reaches of galaxies. The dark matter mystery is one of the deepest conundrums in astronomy. Some researchers think the answer could be massive but invisible, supersymmetric companion particles. Could a supersymmetric payload be the hidden ballast loading down the cosmic craft? Soon the world's most powerful high-energy device could possibly reveal nature's unseen cargo.

Resolving all of these mysteries requires the impact of high-energy collisions monitored carefully by sophisticated detectors to determine the properties of their massive byproducts. Such methods have a long and distinguished history. The story of using collisions to probe the deep structure of matter began a century ago, with gold-foil experiments conducted in 1909. Naturally, the instruments used were far, far simpler back then.

Scientists at that time were trying to explore the inner world of the atom. Little was known about the atomic interior until collisions revealed its secrets. You can't crack open a coconut through the impact of a palm leaf; you need a sturdy mallet applied with vigor and precision. Revealing the atom's structure would require a special kind of sledgehammer and the steadiest of arms to wield it.

3

Striking Gold

Rutherford's Scattering Experiments

Now I know what the atom looks like!

—ERNEST RUTHERFORD, 1911

In a remote farming region of the country the Maoris call Aotearoa, the Land of the Long White Cloud, a young settler was digging potatoes. With mighty aim, the boy broke up the soil and shoveled the crop that would support his family in troubling times. Though he had little chance of striking gold—unlike other parts of New Zealand, his region didn't have much—he was nevertheless destined for a golden future.

Ernest Rutherford, who would become the first to split open the atom, was born to a family of early New Zealand settlers. His grandfather, George Rutherford, a wheelwright from Dundee, Scotland, had come to the Nelson colony on the tip of the South Island to help assemble a sawmill. Once the mill was established,

Ernest Rutherford (1871–1937), the father of nuclear physics.

the elder Rutherford moved his family to the village of Brightwater (now called Spring Grove) south of Nelson in the Wairoa River valley. There, George's son James, a flax maker, married an English settler named Martha, who gave birth to Ernest on August 30, 1871.

Attending school in Nelson and university at Canterbury College in Christchurch, the largest and most English city on the South Island, Rutherford proved diligent and capable. A fellow student described him as a "boyish, frank, simple, and very likable youth, with no precocious genius, but once he saw his goal, he went straight to the central point."[1]

Rutherford's nimble hands could work wonders with any kind of mechanical device. His youthful pursuits would prepare him well for his deft manipulation of atoms and their nuclei. With surgical dexterity, he disassembled clocks, constructed working models of a water mill, and put together a homemade camera that he used to snap pictures. At Canterbury, he became fascinated by the electromagnetic discoveries taking place in Europe, and decided to build his own instruments. Following in the footsteps of Gustav Hertz, he constructed a radio transmitter and receiver—research that would anticipate Guglielmo Marconi's invention of the wireless telegraph. Rutherford showed how radio waves could travel long distances, penetrate walls, and magnetize iron. His clever undertakings would make him an attractive candidate for a new research program in Cambridge, England.

Coincidentally, in the year of Rutherford's birth, a new physical laboratory had been established at Cambridge, with Maxwell its first director. Named after the esteemed physicist Henry Cavendish, who, among other accomplishments, was the first to isolate the element hydrogen, Cavendish Laboratory would become the world's leading center for atomic physics. Its original location was near the center of the venerable university town on a narrow side street called Free School Lane. Maxwell had supervised its construction and planned out its equipment, making it the first laboratory in the world dedicated to physics research. Following Maxwell's death in 1879, another well-known physicist, Lord Rayleigh, had assumed the directorship. Then, in 1884, that mantle passed to the extraordinary leadership of J. J. (Joseph John) Thomson.

An intense intellectual with long, dark hair, wire-framed glasses, and a scruffy mustache, Thomson adroitly presided over a revolution in scientific education that allowed students vastly more opportunities for research. For physics students of earlier times, experimental research was merely the dessert course of a long banquet of mathematical studies—a treat that their tutors would sometimes only grudgingly allow them to partake. After satisfying their requirements with theoretical examinations of mechanics, heat, optics, and so forth, students would perhaps get a chance to sample some of the laboratory apparatus. At Cavendish, with its state-of-the-art equipment, these brief tastes would become a much richer meal unto itself. Thomson was pleased to take advantage of a new program allowing students from other universities to come to Cambridge, perform supervised laboratory research, write up their results in a thesis, and receive a postgraduate degree. Today we are accustomed to research Ph.D.s—it's the bread and butter of academia—but back in the late nineteenth century the concept was novel. Such graduate student assistance would help spark the revolution in physics soon to follow.

The new program began in 1895, with Rutherford one of the first invitees. He was funded through an 1851 scholarship offered to talented young inhabitants of the British Dominion (now the Commonwealth). His move from rural New Zealand to academic Cambridge would prove extraordinary not only for his own career but also for the history of atomic physics.

The moment Rutherford encountered his fate is a matter of legend. Reportedly, his mother received a telegram bearing the good news and brought it out to the potato garden where he was digging. When she told him what he had won, at first he couldn't believe his ears. After the realization sank in, he tossed his spade aside and exclaimed, "That's the last potato I'll dig."[2]

Bringing his homemade radio detector along, Rutherford sailed to London, where he promptly slipped on a banana peel and injured his knee. Fortunately, the country lad had no more missteps as he made his way through the smoky, labyrinthine city. Journeying north, he left the smoke for the fresh air of the English countryside and arrived at the hallowed jumble of colleges and courtyards on the River Cam. There he took up residence at Trinity College, founded in 1546 by King Henry VIII, where the arched Great Gate and legends of Newton's feats tower over nervous entering students. (Cambridge is organized into a number of residential colleges, of which Trinity is the largest.) From Trinity, Cavendish was just a short, pleasant walk away.

Along with Rutherford, the labs of Cambridge were soon filled with research students from around the world. Reveling in the cosmopolitan atmosphere, Thomson invited his young assistants for tea in his office every afternoon. As Thomson recalled, "We discussed almost every subject under the Sun except physics. I did not encourage talking about physics because the meeting was intended as a relaxation . . . and also because the habit of talking 'shop' is very easy to acquire but very hard to cure, and if it is not cured the power of taking part in a general conversation may become atrophied for want of use."[3]

Despite Thomson's efforts to help young researchers lighten up, the pressures at Cambridge must have been intense. "When I come home from researching, I can't keep quiet for a minute and generally get in a rather nervous state," Rutherford once wrote. His solution to his nervousness was to take up pipe smoking, a habit he would maintain for life. "If I took to smoking occasionally," he continued, "it would keep me anchored a bit. . . . Every scientific man ought to smoke as he has to have the patience of a dozen Jobs in research work."[4]

To make matters worse, many of the traditional students viewed the newcomers as interlopers. Taunted by some of his upper-crust colleagues as a yokel from the antipodes, Rutherford bore an extra burden. About one such mocker, he commented, "There is one demonstrator on whose chest I should like to dance a Maori war-dance."[5]

Thomson was a meticulous experimentalist and had been engaged for a time in his own explorations of the properties of electricity. Constructing a clever apparatus, he investigated the combined effects of electric and magnetic fields on what was known as cathode rays: negatively charged beams of electricity passing between negatively and positively charged electrodes (terminals attached by wires to each end of a battery). The negative electrode produces the cathode rays and the positive electrode attracts them.

Electric and magnetic fields affect charges in different ways. Applying an electric field to a moving negative charge creates a force *opposite* to the field's direction. In contrast, a magnetic field generates a force *at right angles* to the field's direction. Also, unlike electric forces, magnetic forces depend on the charges' velocities. Thomson found a way of balancing the electric and magnetic forces in a manner that revealed this speed, which he used to determine the ratio of the charge of the rays to their mass. Making the assumption that these rays bear the same charge as ionized hydrogen, he found the mass of the rays to be about ten thousand times smaller than hydrogen's. In other words, cathode rays consist of elementary particles much, much lighter than atoms. Repeating the experiment numerous times under a variety of conditions, he always got the same results. Thomson called these negatively charged particles corpuscles, but they were later dubbed electrons, a name that has stuck. They offered the first glimpse of an intricate world within the atom.

Initially, Thomson's phenomenal discovery was met with skepticism. As he recalled, "At first there were very few who believed in the existence of these bodies smaller than atoms. I was even told long afterwards by a distinguished physicist who had been present at my lecture at the Royal Institution that he thought I had been 'pulling their legs.' I was not surprised at this, as I had myself come to this explanation with great reluctance, and it was only after I was convinced that the experiment left no escape from it that I published my belief in the existence of bodies smaller than atoms."[6]

Meanwhile, on the other side of the English Channel, the discovery of radioactive decay challenged the notion of atomic permanence. In 1896, Parisian physicist Henri Becquerel scattered uranium salts over a photographic plate wrapped in black paper, and was astonished to find that the plate darkened over time due to mysterious rays produced by the salts. Unlike the X-ray radiation found by Roentgen, Becquerel's rays emerged spontaneously without the need for electrical apparatus. Becquerel found that any type of compound containing uranium gave off these rays, in a rate proportional to the amount of uranium, suggesting that the uranium atoms themselves were producing the radiation.

Similarly working in Paris, Polish-born physicist Marie Curie confirmed Becquerel's findings and, along with her husband, Pierre, extended them to two new elements she discovered: radium and polonium. These elements emitted radiation at a higher rate than uranium and diminished in quantity over time. She coined the term radioactivity to describe the phenomenon of atoms spontaneously breaking down by giving off radiation. For their monumental discovery of the impermanence of atoms through radioactive processes, replacing Dalton's century-old static concept with a more dynamic vision, Becquerel and the Curies would share the 1903 Nobel Prize.

Rutherford followed these developments with great interest. While his mentor Thomson was engaged in discovering the electron, Rutherford concentrated his attention on using radioactive materials as a source for ionizing gases. Somehow the emissions from uranium and other radioactive materials seemed to have the property of knocking the electrical neutrality out of surrounding gases, transforming them into electrically active conductors. The radiation seemed to perform the same function as rubbing dry sticks together and producing a spark.

Radioactivity ignited Rutherford's curiosity as well and launched him on a rigorous investigation of its properties that would revolutionize physics. From a novice keen on developing radio detectors and other electromagnetic devices, he would emerge from his experience an extraordinary experimentalist adept at using radiation to decipher the world of the atom. Using the property that magnetic fields steer differently charged particles along distinct paths, he determined that radioactive materials produce positive and negative kinds of emissions, which he named, respectively, alpha and beta particles. (Beta particles turned out to be simply electrons. Villard discovered gamma radiation, a third, electrically neutral type, shortly after Rutherford's classification.) Magnetic fields cause alpha particles to spiral in one direction and beta in the other—like horses racing in opposite directions around a circular track. Testing the ability of each kind of radiation to be stopped by barriers, he demonstrated that alpha particles are more easily blocked than beta. This suggested that alpha particles are larger in size than beta.

In 1898, in the midst of his studies of radioactive materials, Rutherford decided to take time off for a matter of the heart. He headed briefly to New Zealand to marry his high school sweetheart, Mary Newton. They didn't return to England, however. Married life required a decent salary, he reasoned, so he accepted an offer of a professorship at McGill University in Montreal, Canada, that paid five hundred British pounds per year—respectable at the time and equivalent to about fifty thousand dollars today. The couple sailed to the colder clime, where Rutherford soon resumed his investigations.

At McGill, Rutherford focused on trying to uncloak alpha particles and reveal their true identities. Replicating Thomson's charge-to-mass ratio experiment with alpha particles instead of electrons, he determined their charge—curiously finding it to be precisely the same as helium ions. He began to suspect that the most massive products of radioactivity decay were just mild-mannered helium in disguise.

Just when Rutherford could use some help in unraveling atomic mysteries, a new sleuth arrived in town. In 1900, Frederick Soddy, a chemist from Sussex, England, was appointed to a position at McGill. Learning of Rutherford's work, he offered his expertise, and together they set out to understand the process of radioactivity. They developed a hypothesis that radioactive atoms, such as uranium, radium, and thorium, disintegrated into simpler atoms associated with other chemical elements by releasing alpha particles. Interested in medieval history, when alchemists tried to transform base materials into gold, Soddy recognized that radioactive transmutation could lead to the fulfillment of that dream.

In 1903, soon after Rutherford announced their theory of transmutation, Soddy decided to join forces with a noted expert in helium and other inert gases (neon and so forth), chemist William Ramsay of University College, London. Ramsay and Soddy conducted careful experiments in which they collected the alpha particles produced by decaying radium in a glass tube. Then, after the particles accumulated into a gas, they studied its spectral lines and found them identical to those of helium. Spectral lines are bands of specific frequencies (in the visible spectrum, particular colors) that make up the characteristic signature of an element when it either emits or absorbs light. For the emission spectrum of helium, certain violet, yellow, green, blue-green, and red lines always appear, along with two distinct shades of blue. Ramsay and Soddy found this fingerprint in what they observed, offering proof that alpha particles constitute ionized helium.

Soddy would later coin the term "isotope" to describe when elements exist in two or more distinct forms with different atomic weights. For example, deuterium, or heavy hydrogen, is chemically identical to the standard form, but has approximately twice its atomic weight. Tritium, a radioactive isotope of hydrogen, has about three times the weight of the ordinary variety. It decays into helium-3, a lighter isotope of common helium. In what he called the Displacement Law, Soddy demonstrated how alpha decay causes elements to drop down two spaces on the periodic table, as if sliding backward during a game of snakes and ladders. Beta decay, in contrast, causes a move one space forward, to one of the isotopes of the element in the slot ahead. That's exactly what happens when tritium turns into helium-3 and moves forward in the periodic table.

Suppose you encounter a strange kind of marble dispenser with its contents shielded from view. Sometimes blue marbles pop out of the machine and it flashes once. Sometimes red marbles pop out and it flashes twice. What would you think is inside? You might guess that the interior is an even mixture of red and blue marbles, distributed hither and thither like plums in a pudding.

By 1904, physicists knew that atoms transmuted by producing emissions of different charges and masses, but they didn't know how all of these fit together. Thomson decided to venture a guess that positive and negative particles were distributed evenly—with the latter being much smaller and thereby freer to move. He hoped that the proof of his "plum pudding model" would be in the testing, but alas it would turn out to be plumb wrong—disproven, as fate would have it, by his former protégé from New Zealand.

The next stage of Rutherford's life was arguably his most productive. In 1907, the University of Manchester, the northern English setting of Dalton's explorations, appointed him to a new position as chair of physics. Manchester's gain was a huge loss for McGill. By then, Rutherford had become a commanding presence, "riding the crest of a wave" of his own making—as he once boasted to his biographer (and former student) Arthur Eve.[7] As helmsman, he ran a tight ship—recruiting some of the best young researchers, setting them challenging problems, and dismissing those who fell short. With a booming voice and a propensity toward fits of temper and yelling at equipment during stressful moments, the mustached, pipe-smoking professor could be an intimidating commander indeed. Moments of stress and anger would quickly pass, however, like the blazing sun behind calming puffy clouds, and no one could be friendlier, warmer, or more supportive.

Chaim Weizmann, a Manchester biochemist who would later become the first president of Israel, befriended Rutherford at the time and described him as:

> Youthful, energetic, boisterous; he suggested everything but the scientist. He talked readily and vigorously on every subject under the sun without knowing anything about it. Going down to the refectory for lunch I would hear the loud, friendly voice rolling up the corridor. . . . He was a kindly person, but he did not suffer fools gladly.[8]

Comparing Rutherford to Einstein, whom he also knew well, Weizmann recalled:

> As scientists the two men were strongly contrasting types—Einstein all calculation, Rutherford all experiment. The personal contrast was not less remarkable: Einstein looks like an etherealized body, Rutherford looked like a big, healthy, boisterous New Zealander—which is exactly what he was. But there is no doubt that as an experimenter Rutherford was a genius, one of the greatest. He worked by intuition and whatever he touched turned to gold.[9]

At Manchester, Rutherford had meaty goals: using alpha particles to crack open the atom and reveal its contents. Alpha particles, he realized, would be large enough to make ideal probes of deep atomic structure. In particular, he wanted to test Thomson's plum pudding model and see if each atom's interior was an evenly distributed mix of large positive and small negative chunks. To carry out his project, he was lucky enough to snag two prize catches: a precious supply of radium (for which he had vied with Ramsay) and the valuable services of German physicist Hans Geiger, who had worked for the former physics chair. Rutherford assigned Geiger the task of developing a reliable way of detecting alpha particles.

The method Geiger pioneered—counting sparks that pass between electrodes on a metal tube when incoming alpha particles ionize a gas sealed inside, making it a conductor—became the prototype for what would become his most famous invention: the Geiger counter. Geiger counters rely on the principle that electricity travels around closed loops. Each time a sample emits an alpha particle, electricity rounds the circuit between the electrodes and the conducting gas—producing an audible click. Despite the utility of Geiger's innovation, Rutherford usually relied on a second means of detection: using a screen coated with zinc sulfide, a material that lights up when alpha particles hit it through a process known as scintillation.

In 1908, Rutherford took a break from his research to collect the Nobel Prize in Chemistry for his work with alpha particles. He didn't stay away from the lab for long. Equipped with reliable detection techniques, he developed another project involving Geiger, in conjunction with an extraordinary undergraduate, Ernest Marsden.

Just twenty years old at the time (1909), Marsden had a background curiously parallel to Rutherford's. Like Rutherford, Marsden came from humble roots, with a father in the textile industry. Marsden's dad was a cotton weaver from Lancashire, the local English county. Rutherford started his life in New Zealand and ended up in England; for Marsden it would be the reverse. And each performed vital experimental research while still in their undergraduate years. In Marsden's case, he was just completing his course of study when asked to contribute his talents.

Rutherford recalled the simple query that led to Geiger and Marsden's monumental collaboration. "One day Geiger came to me and said, 'Don't you think that young Marsden . . . ought to begin a small research?' Now I had thought that too, so I said, 'Why not let him see if any alpha particles can be scattered through a large angle?' "[10]

Legendary for posing just the right questions at precisely the right time, Rutherford had a hunch that the possibility of alpha particles scattering backward from a metal would reveal something about the material. Although he was curious to see what would happen, he didn't necessarily expect a positive outcome. But given even the slightest chance of the particles bouncing off a hidden something, he felt it would be a sin not to try.

For certain types of sensitive measurements, particle physicists have to be like nocturnal cats on the prowl; they need to see well in the dark to spot the subtle signs of their prey. That's an area in which younger scientists can have an advantage—not just with better vision but also with patience. No wonder Rutherford and Geiger recruited a twenty-year-old for the alpha particle scattering experiment. Marsden was instructed to cover the windows, making the lab as dark as possible, and then to sit and wait until his pupils were dilated enough to sense every errant speck of light. Only then was he supposed to start taking readings.

Placing plates of various thicknesses and types of metal (lead, platinum, and so forth) near a glass tube filled with a radium compound, Marsden waited for alpha particles to emerge from the tube, hit the plates, and either pass through or bounce off. A zinc sulfide screen, acting as a scintillator, was positioned to record the rates and angles of any alpha particles that happened to reflect. After testing each kind of metal, recording the constellations of sparks his sensitive eyes could see, he shared the data with Geiger. They soon realized that thin sheets of gold offered the highest rate of bounces. Even then, the vast majority of alpha particles passed right through the foil as if it were the skin of a ghost. In the rare cases of reflection, the majority took place at very large angles (ninety degrees or higher), indicating that something hard and focused within the gold was bouncing the alpha particles back.

Glowing with excitement, Geiger ran up to Rutherford. "We have been able to get some of the alpha particles coming backwards!" he reported. Rutherford was absolutely delighted.

"It was quite the most incredible event that has ever happened to me in my life," Rutherford recalled. "It was almost as incredible as if you fired a 15-inch shell at a piece of tissue paper and it came back and hit you."[11]

If Thomson's plum pudding model was correct, then alpha particles impacting a gold foil would be modestly diverted by the gelatinous mixture of charges within the gold atoms and bounce back at fairly small angles. But that's not what Geiger and Marsden found. Like champion sluggers in a baseball game, something within the atoms slammed back the projectiles at large angles only if they were within certain narrow strike zones; otherwise, they continued straight through.

In 1911, Rutherford decided to publish his own alternative to the Thomson model. "I think I can devise an atom much superior to J. J.'s," he informed a colleague.[12] His groundbreaking paper introduced the idea that each atom has a nucleus—a tiny center packed with positive charge and the bulk of the atom's mass. When the alpha particles hit the gold, that's what batted them back, but only in the unlikely chance that they were right on target.

Atoms are almost completely empty space. The nuclei constitute but a minuscule portion of their volume—the rest is unfathomable nothingness. If an atom were the size of Earth, then a cross-section of its nucleus would be roughly the size of a football stadium. Rutherford colorfully described striking a nuclear target as akin to locating a gnat in the Albert Hall, a huge performance venue in London.

Despite their minuteness, nuclei play a critical role for determining the properties of atoms. As Rutherford surmised, the amount of positive charge in the nucleus corresponds to its place in the periodic table—called its atomic number. Starting with one for hydrogen, each atom's nucleus houses the positive equivalent of the charge of the electron multiplied by its atomic number. For example, gold, the seventy-ninth element, has a nucleus charged to the positive equivalent of seventy-nine electrons. Balancing out the central positive charge is the same number of negatively charged electrons—rendering the atom electrically neutral unless it is ionized. These electrons, Rutherford asserted, are scattered in a sphere uniformly distributed around the center.

Rutherford's model was a great conceptual leap, but it left certain questions unanswered. Although it brilliantly explained the Geiger-Marsden scattering results, it didn't address many aspects of what was known about the atom at the time. For example, it didn't account for why the spectral lines of hydrogen, helium, and other atoms have distinct patterns. If electrons in the atom are evenly spaced, how come atomic light spectra are not? And how did Planck's quantum concept and Einstein's photoelectric effect, showing how electrons can exchange energy via discrete bundles of light, fit into the picture?

Fortunately, in the spring of 1912, Rutherford's department welcomed a young visitor from Denmark who would help resolve these issues. Niels Bohr, a freshly minted Ph.D. from Copenhagen with an athletic build and a long face with prominent jowls, arrived at Manchester after half a year with Thomson in Cambridge. Bohr had written to Rutherford asking if he could spend some time learning about radioactivity. He had learned from Thomson about Rutherford's nuclear model and was intrigued about exploring its implications. While performing some calculations about the impact of alpha particles on atoms, Bohr decided to introduce the notion that electrons vibrate with only particular values of energy, multiples of Planck's constant. In a stroke, he forever painted atoms with the variegated coating of quantum theory.

After returning to Copenhagen in the summer of that year, Bohr continued his studies of atomic structure, focusing on the question of why atoms don't spontaneously collapse. Something must prevent the negative electrons from plunging toward the positive nucleus, like a meteorite hurtling toward Earth. In Newtonian physics, a conserved property called angular momentum characterizes the tendency of rotating objects to maintain the same rate of turning. Specifically, mass times velocity times orbital radius tends to remain constant—the reason why ballet dancers twirl faster when they tuck their arms closer to their bodies. Bohr noticed that by requiring an electron's angular momentum to be a multiple of Planck's constant divided by twice the number pi (3.1415 . . .), he forced it to maintain specific orbits and energies. That is, electrons could reside only particular distances from an atom's nucleus, in discrete levels called quantum states.

Bohr's insight led to enormous strides in tackling the question of why atomic spectral lines are arranged in certain patterns. In his model of the atom, electrons neither gain nor lose energy if they maintain the same quantum state—a situation akin to an idealized, absolutely stable planetary orbit. Therefore, superficially, Bohr's picture treats electrons like little "Mercurys," "Venuses," and so forth, revolving around a nuclear "Sun." Instead of gravity acting as the central force, the electrostatic attraction between negative electrons and the positive nucleus does the job. At that point, however, the solar system analogy ends, and Bohr's theory veers on a radically different course. Unlike planets, electrons sometimes "jump" from one quantum state to another, either toward or away from the nucleus. These jumps are instant and spontaneous, resulting in either the loss or gain of a quantum of energy, depending on whether the motion is to a lower or higher energy level. In line with the photoelectric effect, these energy quanta, later called photons or light particles, have frequencies equal to the energy transferred divided by Planck's constant. Thus, the particular lines of color in the emission spectra of hydrogen and other atoms are due to the luminous ballast flung away when electrons enact specific dives—generally the longer the dive, the greater the frequency. Indeed, Bohr's calculations matched up perfectly with the known formulas for the spacings of hydrogen spectral lines—a stunning success for his model.

In the winter of 1913, Bohr wrote to Rutherford with his results and was disappointed to receive a mixed response. Ever the practical thinker, Rutherford found what he saw as a major flaw. He informed Bohr, "There appears to me one grave difficulty in your hypothesis, which I have no doubt you fully realize, namely how does an electron decide what frequency it is going to vibrate at when it passes from one stationary state to another? It seems to me that you have to assume that the electron knows beforehand where it is going to stop."[13]

With his perceptive comment, Rutherford identified one of the principal quandaries involving Bohr's atomic model. How can you predict when an electron will abandon the tranquillity of the state it is in and venture to a new one? How can you determine exactly in which state the electron ends up? Bohr's model couldn't say—and Rutherford was bothered.

Only in 1925 would Rutherford's critique be addressed, and even then the answer was most perplexing. Bohr, by that time, had become the head of his own institute for theoretical physics (now the Niels Bohr Institute) in Copenhagen, where he hosted a stellar array of young researchers. One of the very brightest, German physicist Werner Heisenberg, who studied in Munich and Göttingen, developed a brilliant alternative description of electrons in the atom that, although it didn't explain *why* electrons jumped, could accurately calculate the chances of their doing so.

Heisenberg's "matrix mechanics" introduced a new abstraction to physics that confused many old-timers and revolted some of the prominent physicists who understood its implications—most famously Einstein, who argued vehemently against it. It draped a veil of uncertainty around the atom—and indeed all of nature on that scale or smaller—demonstrating that not all physical properties can be glimpsed at once.

Like many a rebellious youth, Heisenberg began his line of reasoning by abandoning many of the long held suppositions of his elders. Instead of treating the electron as an actual orbiting particle, he reduced it to a mere abstraction: a mathematical state. To represent position, momentum (mass times velocity), and other measurable physical properties, he multiplied the representation state by different quantities. His Ph.D. adviser, Göttingen physicist Max Born, suggested encoding these quantities in arrays called matrices—hence the term "matrix mechanics," also known as quantum mechanics. Equipped with powerful new mathematical tools, Heisenberg felt that he could explore the very depths of the atom. As he recalled, "I had the feeling that, through the surface of atomic phenomena, I was looking at a strangely beautiful interior, and felt almost giddy at the thought that I now had to probe this wealth of mathematical structures nature had so generously spread before me."[14]

In classical Newtonian physics, both position and momentum can be measured at the same time. Not so in quantum mechanics, as Heisenberg cleverly demonstrated. If position and momentum matrices are both applied to a state, the order of their application makes a profound difference. Applying position first and then momentum generally produces a different result than applying momentum first and then position. The situation in which the order of operations matters is called noncommutative—in contrast with commutative forms of arithmetic such as addition and multiplication. While four times two is the same as two times four, position times momentum is not the same as momentum times position. This noncommutativity renders it impossible to know both quantities simultaneously with perfect certainty, a state of affairs Heisenberg later formalized as the uncertainty principle.

In quantum mechanics, Heisenberg's uncertainty principle mandates, for example, that when the position of an electron is ascertained, its momentum goes all fuzzy. Because momentum is proportional to speed, an electron can't tell you where it is and how fast it's going at the same time. Electrons are mercurial creatures indeed, not Mercurial—more like elusive quicksilver than an orbiting planet.

Despite the inherent uncertainty in quantum mechanics, as shown by Heisenberg, it offers accurate predictions of probabilities.

So while it doesn't guarantee that a bet will pay off, at least it tells you the odds. For example, it tells you the chances that an electron will plunge from any given state to another. If the chances are zero, then you know that such a transition is forbidden. Otherwise, it is allowed and you can expect a line in the atom's spectrum with corresponding frequency.

In 1926, physicist Erwin Schrödinger proposed a more tangible alternative version of quantum mechanics, called wave mechanics. In line with a theory proposed by French physicist Louis de Broglie, Schrödinger's version imagines electrons as "matter waves"—akin to light waves, but representing material particles rather than electromagnetic radiation. These wave functions respond to physical forces in a manner described by a relationship called Schrödinger's equation. Subject to the electrostatic attraction of an atomic nucleus, for example, wave functions representing electrons form "clouds" of various shapes, energies, and average distances from the center of the atom. These clouds are not actual arrangements of material, but rather distributions of the likelihood of an electron's being in different points in space.

We can think of these wave formations as akin to the vibrations of a guitar string. Because it is attached on both ends, a plucked guitar string produces what is called a standing wave. Unlike a rolling ocean wave heading toward a beach, a standing wave is constrained to move only up and down. Within such restrictions it can have a number of peaks—one, two, or more—as long as it is a whole number, not a fraction. Wave mechanics identifies the principal quantum number of an electron with the number of peaks of its wave function, offering a natural explanation for why certain states exist and not others.

Much to Heisenberg's chagrin, many of his colleagues favored Schrödinger's depiction over his—perhaps because they were accustomed to models of sound waves, light waves, and kindred phenomena. Matrices seemed too abstract. Fortunately, as the sharp-witted Viennese physicist Wolfgang Pauli proved, Heisenberg's and Schrödinger's descriptions are completely equivalent. Like digital and analog clocks they're equally trustworthy instruments and can be relied on according to taste.

Pauli offered his own critical contribution to quantum mechanics: the concept that no two electrons can be in exactly the same quantum state. His "exclusion principle" inspired two young Dutch researchers, Samuel Goudsmit and George Uhlenbeck, to propose that electrons can exist in two orientations, called spin. Contrary to its name, spin has nothing to do with actual spinning, but rather with an electron's magnetic properties. If we imagine placing an electron in a vertical magnetic field (for instance, directly above a magnetized coil of wire), then the electron's own minimagnet could be aligned in the same direction as the external field, called "spin-up," or in the opposite direction, called "spin-down."

Ambidextrous by nature, an electron normally consists of an equal mixture of spin-up and spin-down states. How can a single particle have two opposite qualities at once? In mundane experience, compass needles can't point north and south at the same time, but the quantum world defies conventional explanation. Until an electron's spin is measured, quantum uncertainty dictates that an electron's spin is ambiguous. Only after a researcher switches on an external magnetic field does an electron reduce into a spin-up or spin-down orientation—a process known as wave function collapse.

If two electrons are paired and one is determined to be spin-up, the other automatically flips spin-down. This switching takes place even if the electrons are widely separated—an intuition-defying effect Einstein called "spooky action at a distance." Because of such unintuitive connections, Einstein thought a deeper, more straightforward theory would someday replace quantum mechanics.

Bohr, on the other hand, embraced contradictions. He reveled in unions of opposites—such as the notion that electrons are waves and particles at the same time, which he called the principle of complementarity. Prone to enigmatic statements, he once said, "A great truth is a truth whose opposite is also a great truth." Appropriately, right in the center of his coat of arms he placed the yin-yang symbol of Taoist contrasts.[15]

Despite their philosophical differences, Einstein shared with Bohr the realization that quantum mechanics matches up incredibly well with experimental data. One sign of Einstein's recognition would be his nomination of Heisenberg and Schrödinger for the Nobel Prize in Physics, which Heisenberg was awarded in 1932 and Schrödinger shared with British quantum physicist Paul Dirac in 1933. (Einstein and Bohr were awardees in 1921 and 1922, respectively.)

Rutherford would remain cautious about quantum theory, continuing to focus his attention mainly on experimental explorations of the atomic nucleus. In 1919, Thomson stepped down as Cavendish professor and director of Cavendish Laboratory, and Rutherford was appointed to that venerable position. During his final year at Manchester and his initial years at Cambridge, he focused on bombarding various nuclei with fast-moving alpha particles. Marsden had noticed that where alpha particles collide with hydrogen gas, even faster, more penetrating particles emerge. These were the nuclei of hydrogen atoms. Rutherford repeated Marsden's experiment, replacing the hydrogen with nitrogen, and much to his astonishment, hydrogen nuclei emerged from that gas too. Striking a fluorescent screen, the scintillations produced by the hydrogen nuclei were so faint and tiny that they could be seen only through a microscope. Yet they offered important evidence that the nitrogen atoms were releasing particles from their cores. As the discovery of radioactivity showed that atoms could transmute on their own, Rutherford's bombardment experiments demonstrated that atoms could be altered artificially as well.

Rutherford coined a name for the positively charged particles found in all nuclei: protons. Some researchers wanted to call these "positive electrons," but he objected, arguing that protons are far more massive than electrons and have little in common. When an actual positive electron was discovered, following predictions by Dirac, it ended up being called the positron.

Positrons provided the first example of what is known as "antimatter": similar to ordinary matter but oppositely charged. Protons, on the other hand, are a key constituent of conventional matter.

A new type of particle detector, called the cloud chamber, aided Rutherford and his group in understanding the paths of particles, such as protons, after they are emitted from target nuclei. While scintillators and Geiger counters could measure the rate of emitted particles, cloud chambers could also capture their behavior as they move through space, leading to an improved understanding of their properties.

Cloud chambers were invented by Scottish physicist Charles Wilson, who noticed during a hike up the mountain Ben Nevis that moist air tends to condense into water droplets in the presence of charged particles such as ions. The charges attract the water molecules and pull them out of the air, offering a vapor trail of electrically active regions. Realizing that the same principle could be used to detect unseen particles, Wilson designed a closed chamber filled with cold, humid air that displayed visible streaks of condensation whenever charged particles passed through—similar to the jet trails etched by airplanes in the sky. These patterns can be photographed, providing a valuable record of what transpires during an experiment.

Although Wilson completed his first working model in 1911, it wasn't until 1924 that cloud chambers came into use in nuclear physics. That year Patrick Blackett, a graduate student in Rutherford's group, used such a device to record the release of protons in the transmutation of nitrogen. His data wonderfully complemented Rutherford's scintillation experiments, presenting irrefutable evidence of artificial nuclear decay.

Protons are not the only inhabitants of nuclei. In another of his legendary successful prognostications, in 1920 Rutherford predicted that nuclei harbor neutral particles along with protons. Twelve years later, Rutherford's student James Chadwick would discover the neutron, similar in mass to the proton but electrically neutral. In a key paper written right after Chadwick's discovery, "On the Structure of Atomic Nuclei," Heisenberg introduced the modern picture of protons and neutrons constituting the cores of all atoms.

The nuclear picture helps explain the different types of radioactivity. Alpha decay occurs when nuclei emit two protons and two neutrons at once—an exceptionally stable combination. Beta decay, on the other hand, derives from neutrons decaying into protons and electrons—with the beta particles comprising the released electrons. As Pauli showed, this couldn't be the whole story because some extra momentum and energy couldn't be accounted for. He predicted the existence of an unseen neutral particle that came to be known as the neutrino. Finally, gamma decay represents the release of energy when a nucleus transforms from a higher- to a lower-energy quantum state. While in alpha and beta decay, the number of protons and neutrons in the nucleus alters, resulting in a different element, in gamma decay that quantity stays the same.

From Rutherford's historic techniques and discoveries, an idea was forged: using elementary particles to probe the natural world on its tiniest scale. Radioactive materials pumping out alpha particles offered a reliable source for early investigations of the nucleus. They were perfect for Geiger and Marsden's scattering experiments that proved that atoms have tiny cores. Yet, as Rutherford came to realize, exploring nuclear properties in a fuller and deeper way would require much higher energy probes. Breaking through the nuclear fortress would take a sturdy battering ram—particles propelled through artificial means to fantastically high velocities. He decided that Cavendish would build a particle accelerator—a project that he recognized would entail a certain amount of theoretical know-how. Fortunately, direct from Stalin's own fortress nation, a whiz kid would slip away and bring his cache of quantum knowledge to Free School Lane.

4

Smashing Successes

The First Accelerators

What we require is an apparatus to give us a potential of the order of 10 million volts which can be safely accommodated in a reasonably sized room and operated by a few kilowatts of power. We require too an exhausted tube capable of withstanding this voltage . . . I see no reason why such a requirement cannot be made practical.

—ERNEST RUTHERFORD (SPEECH AT THE OPENING OF THE METROPOLITAN-VICKERS HIGH TENSION LABORATORY, MANCHESTER, ENGLAND, 1930)

The Soviet People's Commissariat of Education issued its coveted stamp of approval, permitting one of its most brilliant physicists, George Gamow, the opportunity to spend a year at Cavendish. He had almost missed the chance due to an odd medical mix-up. During his clearance check-up, his doctor inadvertently reversed the digits of his blood pressure, flagging him for heart disease. Once that error was cleared up, he got the green light. The Rockefeller Foundation generously offered to support his travel and expenses. Fellowships funded through oil wealth weren't exactly the revolutionary means to success Lenin had anticipated, but the Soviet Union at that time viewed the admission of one of its native sons to the world's premier nuclear physics laboratory as a triumph for its educational system.

It is lucky for the history of accelerators that Gamow managed to make it to England. His theoretical insights would offer the critical recipe for breaking up atomic nuclei and place Cavendish in the forefront of the race to build powerful atom smashers. Thanks in part to his contributions, and to the magnificent experimental work of his colleagues, Cavendish would become for a time the leading center for nuclear research in the world.

The Leningrad-trained physicist arrived in Cambridge in September 1928 and quickly found lodging at a boardinghouse. When a friend visited him soon thereafter he was astonished: "Gamow! How did you manage to get this house?"

At Gamow's perplexed look, his friend pointed to the name of the building. By sheer coincidence it was called the "Kremlin."

Several weeks after joining Cavendish, Gamow experienced one of its director's legendary bursts of temper. One day, without prior explanation, Rutherford called Gamow into his office. Red-faced, he started screaming about a letter he had just received from the Soviet Union. "What the hell do they mean?" he bellowed as he shoved it at Gamow.

Gamow read over the letter. Written in scrawled English, it said:

> Dear Professor Rutherford,
>
> We students of our university physics club elect you our honorary president because you proved that atoms have balls.

After Gamow patiently explained that the Russian word for atomic nucleus is similar to its word for cannonball, and that the letter was probably mistranslated, Rutherford calmed down and had a hearty laugh.[1]

Among the first items Gamow procured at Cambridge were instruments that were ideal for striking spherical projectiles and hurling them toward distant targets. It was a set of golf clubs—par for the course at a collision laboratory. Gamow's instructor in the sport was John Douglas Cockcroft, a young Cavendish researcher and avid golfer.

Born in Todmorden, England, in 1897, Cockcroft took a circuitous path to physics. His father ran a cotton-manufacturing business, but, like Rutherford and Marsden, Cockcroft opted for science over textiles. He began studying mathematics at Manchester University, but then World War I broke out and he joined the British army. Returning to Manchester upon the armistice, he switched to electrical engineering and found a job in the field. Finding that career path personally unfulfilling, he enrolled at St. John's College, Cambridge, and made his way to Rutherford's lab.

In golf, it can be frustrating when there's a hill right in front of the green, occluding the direct path to the hole. In that case, you need a bit of strategy to figure out what club to use and how hard to swing it in order to clear the barrier. Tap the ball with not enough force, and it's liable to fall short.

Cockcroft worked on a problem in nuclear physics that offered a similar kind of challenge. He wanted to hurl particles toward nuclear targets with the goal of exciting them to higher energy levels and possibly breaking them into subcomponents. If smashing them together broke them up, seeing what came out would allow him and his colleagues to learn things about the makeup of an atom they couldn't learn in any other way. Blocking the path to the nucleus, however, was the barrier caused by the mutual electrostatic repulsion of positively charged particles and the positive nuclear charge. They naturally push each other apart—a formidable obstacle to overcome—like the north poles of bar magnets resisting being brought together, only far stronger.

Gamow knew just how to handle the issue theoretically. Plugging the parameters corresponding to protons and alpha particles (the particles radioactive atoms such as uranium give off) into his "quantum tunneling" formula, Gamow discovered that the former would need sixteen times less initial energy than the latter for the same probability of penetration. The choice was clear: protons offer much more economical projectiles. If protons could be induced to move fast enough, a few might pass through the force barrier around an atom and smash into its nucleus. What exactly would happen once they reached their targets was unknown, but, convinced by Gamow, Rutherford decided it would be worthwhile to try. It would be the one major step Rutherford took that was driven by theoretical predictions.

Already involved in planning the details of an atom smasher was an adept young experimentalist, Ernest Thomas Sinton Walton. Born in Dungaravan, Ireland, in 1903, he was the son of a traveling Methodist preacher. In 1915, Walton enrolled in a Methodist boarding school where he excelled in the sciences. Following his graduation in 1922, he became a student at Trinity College in Dublin, where he received a master's degree in 1927. Upon being awarded an Exhibition of 1851 scholarship to Cambridge, he joined the group at Cavendish and soon became one of Rutherford's trusted assistants.

In late 1928, Walton came across an extraordinarily innovative research paper by Norwegian engineer Rolf Wideröe that described his attempts to accelerate particles by means of a device called the ray transformer. Wideröe's mechanism combined several basic concepts in electromagnetic theory. It starts with the idea of an electromagnetic coil: a current-carrying wire wrapped in a loop that produces a magnetic field in its vicinity. If the wire has a changing current, then the magnetic field changes over time. Then, according to Faraday's law of induction, the changing magnetic field produces a second current in any wire that happens to be nearby. If that second wire is in a loop too, the setup is known as a transformer—a familiar system for transferring power from one wire to another. In a way, it's analogous to the spinning of a bicycle's pedals giving rise to the turning of its wheels—with the chain representing the varying magnetic field connecting the two.

Wideröe's principal innovation was to replace the second wire with electrons accelerated through a vacuum-filled ring. These electrons would be removed from atoms and propelled through space by what is called the electromotive force produced by the changing magnetic field. To keep the electrons moving in a loop, like race cars on a circular track, he envisioned a central magnet that would steer them round and round. Unfortunately, in trials of his machine at Aachen University in Germany, he found that "islands of electrons" built up in the tube, sapping the revolving electrons' energy. For some reason, the magnet couldn't keep the electrons moving smoothly, though he couldn't figure out why. The best Wideröe could manage given the turbulence was to get the electrons to circle around the loop one and a half times.

Frustrated by the problems with the circular track, Wideröe finally decided to abandon the project and turn to a different scheme. Borrowing a concept he found in a 1924 article by Swedish physicist Gustav Ising, he pursued the idea of a linear accelerator and built a small prototype, about one yard long. Rather than a ring, it used a pair of "drift tubes" (straight, isolated, vacuum-filled pipes) in which particles would be sped up by successive "kicks" of an electric field. These boosts were arranged rather cleverly—allowing the particles to be lifted up to higher speeds twice by use of the same voltage difference—something like the continuously ascending stairways in some of Escher's paintings. Just when the particles seemed to reach the top, there was more to climb.

Voltage, electric potential energy per charge, is a measure of how easy it is for particles of specific mass and charge to accelerate from one place to another; the higher the voltage difference, the greater the acceleration, all other factors being equal. In other words, voltage is a measure of how steep that staircase is—and how much of a boost it gives.

Particles began their journeys through a drift tube with high voltage (twenty-five thousand volts) at the entrance point and low voltage at the exit. This voltage difference caused them to speed up. When the particles were halfway through and already moving quickly, Wideröe tricked them by reversing the voltage difference, setting the formerly low voltage to high. Because they were already moving at high speeds it was too late for them to turn back. They rushed through the tail end of the first tube, across a gap, and then on to the start of a second tube, where the same voltage difference (once again, due to an initially high voltage and a final low voltage) accelerated them even farther. Because it used the same voltage difference twice, Wideröe's method doubled the impact of the boost, enabling a lower voltage source than otherwise required.

At the end of the second tube, Wideröe placed a photographic plate to record the streaks produced by the high-velocity particles as they impacted. Experimenting with potassium and sodium ions as the projectiles, he was able to run them through his device. The ions were made by stripping atoms of their outer electrons. The positively charged ions were then compelled by the voltage differences to accelerate through the tubes before hitting the plate. After collecting enough data, Wideröe incorporated his findings into a doctoral thesis for Aachen University. The thesis was published in a journal his Ph.D. adviser edited.

Fascinated by Wideröe's work, in December 1928, Walton proposed to Rutherford the idea of building a linear accelerator at Cavendish. Rutherford was keen to devise such a device that could look inside one of the lighter elements, such as lithium. (Lithium is the third element in the periodic table after hydrogen and helium and its atom is now known to have three protons and four neutrons in its core.) The next month, Gamow gave a talk that presented his barrier penetration formula to the group. Cockcroft was eager to apply this formula to the issue of penetrating the lithium nucleus with protons. Estimates showed that it would take several hundred thousand electron volts to do the job. By human standards, even 1 MeV (one million electron volts) is an extraordinarily tiny amount of energy—approximately one billionth of a billionth of a single dietary calorie (technically, a kilocalorie). Elementary particles obviously don't have to worry about slimming down—however for them it's quite an energizing burst!

Upon hearing these results, Rutherford called Cockcroft and Walton into his office. "Build me a one-million-electron volt accelerator," he instructed. "We will crack the lithium atom open without any trouble."[2]

Soon Cockcroft and Walton were hard at work building a linear accelerator that they would locate, when it was complete, in a converted lecture hall. They rigged up a straight tube with a specially designed high-voltage power supply, now known as the Cockcroft-Walton generator. It included a mechanism known as a voltage multiplying circuit that included four high-voltage generators stacked in a ladderlike formation twelve feet high. Capacitors (charge-storing devices) in the circuit helped

A Cockcroft-Walton generator, one of the earliest types of accelerator. This example is retired and located in the garden of Microcosm, CERN's science museum.

boost a relatively modest input voltage to an overall voltage of up to seven hundred thousand. Propelled by this high voltage, protons would be accelerated by the electric forces through an evacuated tube and collide with nuclear targets on the other end—with any disintegrations recorded as sparks on a fluorescent screen placed inside the vacuum.

In 1931, Walton received his Ph.D. from Cambridge. With the Cavendish accelerator on the brink of completion, Rutherford certainly couldn't afford to lose one of its principal architects. Walton was appointed Clerk Maxwell Research Scholar, a position he would hold for an additional three years while continuing to work with Cockcroft and Rutherford.

Cavendish was far from the only player in the race to split the atom. Physicists around the world were well aware of what Rutherford was trying to do and hoped to unlock the nucleus themselves with their own powerful atom smashers. Aside from scientific interest, another motivation that would become increasingly important was anticipation of colossal energy locked inside the atomic core. Einstein's famous equation for the equi-valence of energy and mass, $E = mc^2$, indicated that if any mass were lost during a nuclear disintegration it would be converted into energy—and this could be formidable. In 1904, even before Einstein's finding, Rutherford had written, "If it were ever possible to control at will the rate of disintegration of the radio elements, an enormous amount of energy could be obtained from a small amount of matter."[3] (In 1933, he would qualify this statement when, in a prediction uncharacteristically off the mark, he expressed the opinion that atomic power could never be controlled in a way that would be commercially viable.)

A highly innovative thinker who would become a key participant in the development of nuclear energy was the Hungarian physicist Leo Szilard. In December 1928, while living in Berlin, Szilard took out a patent for his own concept of a linear accelerator. Like Ising and Wideröe, Szilard envisioned an oscillating (direction switching) electric field that would prod charges along. In his patent application, titled "Acceleration of Corpuscles," he described a way of lining up charged ions so they ride the crest of a traveling wave forcing them to move ever faster:

> With our arrangement, the electric field can be conceived of as a combination of an electric field in accelerated motion from left to right and an electric field in decelerated motion from right to left. The device is operated in such a way that the velocity of the accelerated ion equals, at each point, the local velocity of the field moving left to right.[4]

Curiously, Szilard never pursued his design. He developed patent applications for two other accelerator schemes that he similarly never followed up on. History does not record whether or not his patent applications were even accepted—conceivably the patent officers were aware of the earlier papers by Ising and Wideröe.

Around the same time as Cockcroft and Walton began to build their apparatus, American physicist Robert Jemison Van de Graaff developed a simple but powerful accelerator model that, because of its compactness and mobility, would become the nuclear physics workhorse for many years. Born in Tuscaloosa, Alabama, in 1901, Van de Graaff began his career on a practical track. After receiving B.S. and M.S. degrees in mechanical engineering from the University of Alabama, he worked for a year at the Alabama Power Company. He could well have remained in the electrical industry, but Europe beckoned, and in 1924, he moved to Paris to study at the Sorbonne. The great Marie Curie herself taught him about radiation—acquainting him in her lectures with the mysteries of nuclear decay. His savvy won him a Rhodes scholarship, enabling him to continue his studies at Oxford. There he learned about Rutherford's nuclear experiments and the quest to accelerate particles to high velocities. Oxford awarded him a Ph.D. in physics in 1928.

In 1929, Princeton University appointed Van de Graaff as a national research fellow at the Palmer Physics Laboratory, the center of its experimental program. He soon designed and built a prototype for a novel kind of electrostatic generator that could build up enormous amounts of charge and deliver colossal jolts. Its basic idea is to deliver a continuous stream of charge from a power source to a metal sphere using a swift, insulated conveyor belt. Van de Graaff constructed his original device using a silk ribbon and a tin can; later he upgraded to other materials. Near the bottom of the belt, a sharp, energized comb connected to the power source ionizes its immediate surroundings, delivering charge to the belt. Whisked upward, the charge clings to the belt until another comb at the top scrapes it off and it passes to the sphere. A pressurized gas blankets the entire generator, creating an insulated cocoon that allows more and more charge to build up on the sphere.

Using the Van de Graaff generator as an accelerator involves placing a particle source (a radioactive material or an ion source, for example) near the opening of a hollow tube, each situated within the sphere. The voltage difference between the sphere and the ground serves to propel the particles through the tube at high speeds. These projectiles can be directed toward a target at the other end.

Van de Graaff worked continuously at Princeton and later at MIT to increase the maximum voltage of his generators. While his prototype could muster up to eighty thousand volts, an updated model he presented at the American Institute of Physics's inaugural banquet in 1931 stunned the dining guests (fortunately not literally) by producing more than one million volts. A much larger machine he assembled on flatbed railroad cars in a converted aircraft hangar in South Dartmouth, Massachusetts, consisted of twin insulated columns, each twenty-five feet high, capped with fifteen-foot-diameter conducting spheres made of shiny aluminum. Its colossal power inspired the *New York Times* headline on November 29, 1933, "Man Hurls Bolt of 7,000,000 volts."[5]

According to Greek mythology, Prometheus stole the secret of fire from the gods, offering humanity the sacred knowledge of how to create sparks, kindle wood, light torches, and the like. Yet even after that security breach, mighty Zeus reserved the right to hurl thunderbolts at his foes, illuminating the heavens with his terrifying power. Through Van de Graaff's generator, even something like the awe-inspiring vision of lightning became scientifically reproducible, albeit on a smaller scale, ushering in a new Promethean age in which colossal energies became available for humankind's use. Given such newly realized powers, perhaps it is not surprising that many horror films of the day, such as *Frankenstein* (1931) and *Bride of Frankenstein* (1935) most notably, offered sinister images of gargantuan laboratories spawning monsters electrified by means of colossal generators.

Why rely on expensive generators for artificial thunderbolts, if the real thing is available in the skies for free? Indeed, though lightning is, of course, highly unpredictable and extremely dangerous, several physicists of that era explored the possibility of using lightning itself to accelerate particles. During the summers of 1927 and 1928, University of Berlin researchers Arno Brasch, Fritz Lange, and Kurt Urban rigged up an antenna more than a third of a mile across between two adjacent peaks in the southern Swiss Alps near the Italian border. They hung a metallic sphere from the antenna and wired another sphere to the ground to measure the voltage difference between the two conductors during thunderstorms. During one lightning strike, more than fifteen million volts passed through the device, according to the researchers' estimates. Sadly, during their investigations Urban was killed. The two survivors returned to Berlin to test discharge tubes with the potential to withstand high voltages. Brasch and Lange published their results in 1931.[6]

Lightning strikes, even of the artificial kind, are usually one-time-only affairs. Whenever great quantities of charge build up, it creates a huge voltage drop that maintains itself as long as the collected charge has nowhere else to go (if the apparatus is isolated or insulated, for example). Like cliff divers, particles plunging down the steep potential difference experience a force that accelerates them. But once they reach the ground, that's it—end of story.

However, as Wideröe pointed out in his "ray transformer" proposal, particles forced to travel in a ring, instead of a straight line, could be accelerated repeatedly each time they rounded the loop, building up to higher and higher energies. Although, after his initial experiments failed, Wideröe ceased working on the idea of a circular accelerator, his article inspired an extraordinary American physicist, Ernest Orlando Lawrence, to pursue this vital approach.

Lawrence was born in the prairie town of Canton, South Dakota, in 1901. His parents, Carl and Gunda, were both schoolteachers of Norwegian descent. Carl was the local superintendent of schools and also taught history and other subjects in the high school; Gerda instructed in mathematics. Ernest was a cheerful baby, whom the neighbors across the street, the Tuve family, contrasted with their own colicky, crying boy born six weeks earlier, Merle.

Practically from birth Ernest and Merle were best friends. They would pull various pranks together, such as once dumping rubbish on another neighbor's porch. The neighbor happened to be home and snatched Merle, before he escaped, through a hole in her fence. Meanwhile, Ernest managed to get away. They shared a code of honesty and tried not to fib, even when they were mischievous.

When the boys were only eight years old, they became interested in electrical devices. Practically all of their waking hours, aside from school and chores, were spent hooking up crude batteries into circuits, connecting these power sources to bells, buzzers, and motors, and testing which combinations worked best.

Tall and lanky, Lawrence earned the childhood nickname "Skinny," which he didn't seem to mind. His interests were as narrow as his build. Aside from tennis, he was little interested in sports and participated grudgingly in athletic activities, mainly when prodded by his father. Nor, as a teenager, was he much inclined to go on dates and other social events. Rather, he buried himself in his studies so that he could graduate from high school a year early, while continuing to spend his free time, along with Tuve, assembling various mechanical and electrical apparatuses. To earn money for switches, tubes, and other radio equipment, he spent one summer working on an area farm—a job he detested. The farmer he worked for had a low opinion of his skills, complaining, "He can't farm worth a nickel."[7]

Despite his limitations in many areas beyond science, Lawrence's single-mindedness would prove a great strength. Like a magnifying glass on dry wood, whatever topics Lawrence's bright blue eyes did focus on would be set ablaze with his extraordinary energy and intuitive understanding. One of the first to recognize his talents was Lewis Akeley, dean of electrical engineering at the University of South Dakota, where he completed his undergraduate studies. Lawrence had transferred there in 1919 from St. Olaf College in Minnesota with the goal of preparing for a career in medicine, but Akeley steered him toward physics. Akeley was so impressed by Lawrence's phenomenal knowledge of wireless communication that he decided to experiment with a new teaching arrangement. For senior seminar he asked Lawrence, the only upper-level physics major, to prepare and deliver the lectures himself. While Lawrence was speaking, Akeley sat smiling as an audience of one, marveling that he was lucky enough to get to know perhaps the next Michael Faraday.

Meanwhile, Tuve was at the University of Minnesota and persuaded Lawrence to join its graduate program in physics. There, Lawrence found a new mentor, English-born physicist W. F. G. Swann, from whom he learned about the latest questions in quantum physics. Swann was a bit of a restless soul, an accomplished cellist as well as a researcher, who hated stodginess and valued creative thinking. Unhappy in Minnesota, he moved to Chicago and then to Yale, inspiring Lawrence to follow. Lawrence received his Ph.D. from Yale in 1925 and continued for three more years as national research fellow, working on new methods for determining Planck's constant and the charge-to-mass ratio of the electron. Along with another research fellow, Jesse Beams, he developed a highly acclaimed way of measuring extremely short time intervals in atomic processes. They showed that the photoelectric effect, in which light releases electrons from metals, takes place in less than three billionths of a second—lending support to the idea that quantum events are instantaneous.[8]

With his Ph.D. in hand, Lawrence finally found time to socialize, albeit in a manner unusual for a postdoctoral researcher. The daughter of the dean of the medical school, Mary Kimberly "Molly" Blumer, who was then only sixteen years old, needed a date for her high school prom. Word got out, and Lawrence agreed to be her escort. He was impressed with her quiet thoughtfulness, and after the prom he asked her if he could see her again. Politely, she said he could stop by, even though at that point in her life she understandably felt awkward being wooed by a man nine years her senior. Each time he visited her house, she would take any measure not to be alone with him—making sure her sisters accompanied them at all times. Sometimes she would even hide out in a family fishing boat in the Long Island Sound and refuse to return to shore. It is a tribute to his perseverance that they would eventually get married.

Lawrence was courted in a different way by a rising star in the academic world, the University of California in Berkeley. Berkeley offered him an assistant professorship. When he turned it down in favor of remaining at Yale, Berkeley raised its offer to an associate professorship—a rank usually reserved for more seasoned faculty. Lawrence then decided to accept, thinking Berkeley would offer quicker advancement and a greater chance of working directly with graduate students. Some of his stuffier colleagues at Yale were taken aback that he would even consider switching to a non–Ivy League institution. "The Yale ego is really amusing," Lawrence wrote to a friend. "The idea is too prevalent that Yale brings honor to a man and that a man cannot bring honor to Yale."[9]

In August 1928, Lawrence drove out west in an REO Flying Cloud Coupe to assume his new position. After crossing the American heartland and reaching the rolling Berkeley hills, he took time to admire the beauty of the Bay Area and the exciting cultural jumble of San Francisco. The campus, dominated by a Venetian-style bell tower, offered a different kind of splendor. Although rooted in European architectural themes, it seemed refreshingly bright and modern—a far cry from pompous East Coast tradition.

Blessed with ample space and support for his work, he resumed his studies of the precise timing of atomic processes. Then, barely seven months after he started, his research took an unforeseen turn. Around April Fools' Day of a year devastating for investors but auspicious for high-energy physics, Lawrence was sitting in the Berkeley library browsing through journals. Wideröe's article leapt to his attention as if it were spring-loaded. It was the diagrams, not the words, that he noticed at first—sketches of electrodes and tubes arranged to propel particles.

Of Wideröe's two accelerator designs, the linear dual-tube arrangement and the ringed "ray transformer" scheme, Lawrence found the latter more appealing. Lawrence realized immediately that a straight-tube setup would be limiting, providing only a few kicks before particles reached their targets. By curving the tubes into semicircles with electrifying gaps in between, and bending particle paths with a central magnet, he saw that he could jolt the particles again and again. He noted the fortuitous coincidence in magnetism that for a given particle steered in a circular loop by a magnetic field, if the field is constant the ratio of the particle's velocity to its orbital radius—a quantity known as angular velocity—similarly remains the same, even if the particle speeds up. Because angular velocity represents the rate by which an object travels around a circle, if it is constant then the object passes the same point in equal intervals of time—like a racehorse passing a grandstand precisely once a minute. This regularity, Lawrence, determined, would ensure that a voltage boost peaked at regular intervals (oscillating in the same rhythm as the orbits) could accelerate particles around a ring to higher and higher energies until they reached the level needed to penetrate a target nucleus. Lawrence realized the problem with Wideröe's design, and his pileup of electrons, was all in timing the voltage boosts.

Lawrence shared his design with Berkeley mathematician Donald Shane, who verified that the equations checked out. When Shane inquired, "What are you going to do with it?" Lawrence excitedly replied, "I'm going to bombard and break up atoms!"

The following day, his exuberance grew even greater when additional calculations confirmed that particles in his planned accelerator could continue to circle faster and faster no matter how far they spiraled outward from the center of the ring. As he strutted through campus like a proud peacock, a colleague's wife distinctly heard him exclaim, "I'm going to be famous!"[10]

As he was accustomed to during childhood, Lawrence couldn't wait to run his idea past Tuve, who was then working at the Carnegie Institution in Washington. Tuve was dubious about the scheme's practicality. Ironically, the best friends were becoming rivals—following separate tracks in the race to split the nucleus. Along with Gregory Breit and Lawrence Hafstad, Tuve was involved in efforts to crank Tesla coils—paired wound coils for which lower voltage in one induces high voltage in the second—up to ultrahigh energies. However, these devices were extremely hard to insulate and wasted a lot of energy. After Van de Graaff developed his high-voltage generators, Tuve recognized their promise and began constructing his own.

Because of doubts expressed by Tuve and other colleagues, Lawrence was at first hesitant to try out his concept, which he initially called the magnetic resonance accelerator and later became known as the cyclotron. It took some prodding by a noted scientist to get him going. Around Christmas 1929, he sat down for some bootleg wine (it was Prohibition) with German physicist Otto Stern—who was visiting the United States at the time—and outlined his scheme. Stern became excited and urged Lawrence to turn his design into reality. "Ernest, don't just talk any more," he urged. "You must . . . get to work on that."[11]

Placing himself in direct competition with Cockcroft, Walton, Van der Graaff, Tuve, and other nuclear researchers, Lawrence hadn't a moment to spare to get his accelerator up and running. He took his first Ph.D. student, Niels Edlefsen, aside and inquired, "Now about this crazy idea of mine we've discussed. So simple I can't understand why someone hasn't tried it. Can you see anything wrong with it?"

Edlefsen responded that the idea was sound. "Good!" said Lawrence. "Let's go to work. You line up what we need right away."[12]

Under Lawrence's supervision, Edlefsen assembled a hodgepodge of materials found around the lab into a working prototype. A round copper box, cut in half, served as the two

A thirty-seven-inch early cyclotron at the Radiation Laboratory, now the Lawrence Berkeley National Laboratory.

electrodes—wired to a radio-frequency oscillator that offered a cyclic voltage boost. Edlefsen encased the device in glass and centered it between the four-inch poles of a guiding electromagnet. Finally, he sealed all of the connections with sticky wax. Completed in early 1930, it wasn't very elegant. Nevertheless it was sufficient, after some tinkering, to get protons circulating—much to Lawrence's delight.

For energies high enough to break through nuclear barriers, Lawrence realized that he needed a bigger machine with a more powerful magnet. Fortunately an industrial executive, who also taught at the university, offered him an eighty-ton magnet that had been gathering dust in a warehouse about fifty miles south in Palo Alto. Built for radio transmission, it had been rendered obsolete by technological advances.

The generous donation prodded Lawrence to find a more spacious setting suitable for a much larger accelerator. He got lucky again; in 1931, an old building on campus was about to be demolished and he was given permission to use it. Christened the Radiation Laboratory (and nicknamed the "Rad Lab"), it and its successor buildings would serve for decades as his dedicated center for research. It would eventually be renamed in honor of its founder and is now known as the Lawrence Berkeley National Laboratory.

Another pressing issue was how to move the colossal magnet to the lab. When yet another donor offered him funds to that effect, Lawrence scored a triple play in the tight budgetary age of the Great Depression. He finally had ample space and equipment to construct a powerful machine.

In 1932, a banner year for nuclear physics, remarkable experiments around the world cast powerful spotlights on the murky inner workings of atoms. At Columbia University, chemist Harold Urey discovered deuterium, an isotope of hydrogen with approximately twice the mass of the standard version. James Chadwick's identification of the neutron, found through meticulous observations at Cavendish, explained why deuterium is twice as massive as its similarly charged brother: the heavier isotope is bloated with extra neutrons. Speculations arose as to whether neutrons are particles in their own right, or alternatively protons and electrons somehow clumped together to make an electrically neutral particle.

A couple of different theories had been bandied about, and only experimentation could tell which of the theorists has guessed right. For example, beta decay is when a radioactive substance gives off electrons. Those electrons, some suggested, must be coming from neutrons breaking into protons and electrons. (We now know that it is the weak interaction that is mediating a transformation involving the quarks that form protons and neutrons, along with the electron and a neutrino.)

Carl Anderson's discovery of the positron offered another possible explanation for the relationship between neutrons and protons. He found the positron in cloud chamber photographs taken at Caltech of positively charged cosmic rays (radiation from space passing through Earth's atmosphere) with the same mass as the electron. We now know a positron is the antimatter version of an electron, but at the time, Anderson wondered if the neutron is fundamental, and the proton an amalgamation of a neutron and a positron. Testing these alternatives would require precise measurements of the masses of protons and neutrons, to see if one was sufficiently heavier than the other to accommodate an electron or positron. (Indeed, as we now know, the neutron is heavier, but is composed of quarks, not protons and electrons.)

While Lawrence, along with Wisconsin-born graduate student M. (Milton) Stanley Livingston, toiled on the larger cyclotron, word came of victory in the race to split the lithium nucleus. The first to reach the finish line were Cockcroft and Walton, using the Cavendish linear accelerator. Walton recalled the moment of discovery when they finally bombarded the lithium target with protons and observed the stunning results:

> On the morning of April 14, 1932, I carried out the usual conditioning of the apparatus. When the voltage had risen

> to about 400,000 volts, I decided to have a look through the microscope which was focused on the fluorescent screen. By crawling on my hands and knees to avoid the high voltage, I was able to reach the bottom of the accelerating tube. To my delight, I saw tiny flashes of light looking just like the scintillations produced by alpha particles which I had read about in books but which I had never previously seen.[13]

After observing what surely looked like the decay of lithium, Walton called Cockcroft into the lab, who agreed with that explanation. Then they invited Rutherford to crawl into the chamber and check out the scintillations himself. They turned off the voltage, and he ducked inside. When Rutherford came out, he said:

> Those scintillations look mighty like alpha particle ones and I ought to be able to recognize an alpha particle scintillation when I see one. I was in at the birth of the alpha particle and I have been observing them ever since.[14]

Uncharacteristically, Rutherford asked Cockcroft and Walton to keep the news a "dead secret" until they could conduct more measurements. As Walton explained in a letter to his girlfriend, Freda Wilson (whom he would marry in 1934):

> He [Rutherford] suggested this course because he was afraid that the news would spread like wild fire through the physics labs of the world and it was important that no lurid accounts should appear in the daily papers etc. before we had published our own account of it.[15]

Cockcroft and Walton ran the experiment further times using a cloud chamber to record the alpha particle tracks. (Recall that a cloud chamber is a box full of vapor for which charged particles passing through create a visible misty trail.) Calculating the masses before and after the collision, they confirmed that each lithium nucleus, with three protons and four neutrons, had been cajoled by an extra proton to break up into two alpha particles, each of two protons and two neutrons. They'd literally cut the lithium ions in half!

Moreover, the energy released during each hit corresponded precisely to the mass difference between the initial and final states, times the speed of light squared. Their experiment confirmed Einstein's famous formula. Satisfied with the accuracy and importance of their results, they published their findings in the prestigious journal *Nature*. For their exemplary work, Cockcroft and Walton would share the 1951 Nobel Prize for physics.

The news from Cambridge didn't dampen Lawrence's spirits. He had much to celebrate. For one thing, he had just married Molly and was on his honeymoon. His dogged persistence in romance as well as in science had finally paid off. The coy young woman had grown to love her awkward but accomplished suitor. They would have a large family together—four girls and two boys.

Another cause for Lawrence's optimism was his strong conviction that he was at the forefront of a new scientific era. Ultimately, he realized, cyclotrons could yield much higher energies than linear accelerators could muster, and would thereby be essential for future probes of the nucleus. He wasted no time in confirming Cockcroft and Walton's lithium results with an eleven-inch cyclotron. The larger device in the Rad Lab with the eighty-ton magnet was still under completion. When it was ready in March 1933, Lawrence bombarded lithium with protons and generated a bounty of highly energetic alpha particles—ricocheting back with impressive range. He also struck a variety of elements with deuterons, producing protons of Olympian stamina—some sprinting up to fifteen inches. By that point, he was more than ready to share his results with the physics community at large.

The Seventh Solvay Conference, held in Brussels during the last week of October 1933, was a milestone for discussion of the remarkable advances in nuclear physics. Among the scientific luminaries present were quantum pioneers Bohr, de Broglie, Pauli, Dirac, Heisenberg, and Schrödinger. The Parisian contingent included Marie Curie, along with her daughter and son-in-law, Irene Joliot-Curie and Frédéric Joliot, each an esteemed nuclear chemist and future Nobel laureate.

From Russia came Gamow—the start of his permanent exile, as it turned out. Two years earlier, he had returned to his homeland by way of Copenhagen. Unhappy living under Stalin's iron thumb, he and his wife had attempted to escape across the Black Sea to Turkey but had been foiled by foul weather. Remarkably, an invitation by Bohr allowed both of them to slip into Belgium, where Gamow announced to his surprised host that they would never go back.

The Cavendish contribution to the meeting was impressive. Headed by Rutherford, it included Cockcroft, Walton, Chadwick, and Blackett. Finally, though Lawrence was the lone American attendee, his presence was vital, in as much that cyclotrons represented the future of nuclear exploration and that the United States would for decades be the principal testing ground for such devices.

Cockcroft delivered the conference's first talk, "The Disintegration of Elements by Accelerated Protons." Listening eagerly to his every word was Lawrence, keen to demonstrate the superiority of the cyclotron in handling the job. Perusing Cockcroft's handout, Lawrence noted a statement that "only small currents are possible" from the cyclotron and emphatically crossed it out. In the margin he wrote, "Not true," expressing his clear impatience with Cockcroft's assertions.[16]

When it came time for discussion, Lawrence was quick to respond. He presented an account of his own device and argued that it offered the best way forward to explore the nucleus. He also offered his own estimation of the mass of the neutron—controversially, much lower than Chadwick's value. Further experiments conducted by Tuve later that year would demonstrate that Lawrence was wrong; a mistake he would frankly acknowledge. The neutron turned out to be slightly bulkier than the proton.

After Solvay, Lawrence traveled to England and spent a pleasant couple of days at Cavendish. Rutherford warmly welcomed him and led him on a personal tour. After some heated discussions about lithium bombardment results, Rutherford said of Lawrence, "He's a brash young man but he'll learn."[17]

Lawrence tried to convince Rutherford to build a cyclotron at Cavendish. Chadwick and Cockcroft joined in the chorus, arguing that it was the only way for the lab to remain competitive. Rutherford would not budge. He had a preference for homemade equipment and was reluctant to import another group's idea. Moreover, he disliked trolling for funds and knew a cyclotron would be expensive.

Rutherford's reluctance cost him dearly. In 1935, Chadwick, frustrated with the lack of progress, departed for a position at the University of Liverpool where he began to solicit funds for a cyclotron. On a visit to Cambridge in the summer of 1936, he and his former mentor were barely on speaking terms. Around the same time, Australian-born physicist Mark Oliphant, another of Rutherford's protégés, was offered a position at the University of Birmingham, which he would assume the following year. Pressed by the loss of some of his top researchers, an embittered Rutherford finally agreed to let Cockcroft construct a cyclotron at Cambridge.

While Rutherford hedged, Lawrence was busy collecting funds to build an even larger cyclotron at Berkeley. Tremendously successful at fund-raising, he had no trouble continuing to expand the Radiation Laboratory's work. Oliphant, who would visit there and get to know Lawrence as well as Rutherford, explained the difference in their styles: "The Cavendish laboratory, under Rutherford and his predecessors, was always short of money. Rutherford had no flair and no inclination for raising funds. . . . Lawrence, on the other hand, had shrewd business sense and was adept at raising funds for the work of his laboratory."

Oliphant pointed out that Lawrence, who originally was on a premed track at university, had the savvy to foresee the medical applications of cyclotrons and how these could be used to draw funding. In a 1935 letter to Bohr, Lawrence wrote, "As you know, it is so much easier to get funds for medical research."

Unlike Rutherford, who suggested and personally supervised almost every experiment his lab undertook, Lawrence liked to delegate authority. He had exemplary management skills that impressed his benefactors in government and industry and enabled his lab to expand. As Oliphant noted:

> His direct approach, his self-confidence, the quality and high achievement of his colleagues, and the great momentum of the researchers under his direction bred confidence in those from whom the money came. His judgment was good, both of men and of the projects they wished to undertake, and he showed a rare ability to utilize to the full the diverse skills and experiences of the various members of his staff. He became the prototype of the director of the large modern laboratory, the costs of which rose to undreamt of magnitude, his managerial skill resulting in dividends of important scientific knowledge fully justifying the expenditure.[18]

On October 19, 1937, Rutherford died of a strangulated hernia. Having been raised to peerage six years earlier, he was buried with the honors accorded his position as "Right Honourable Lord of Nelson." His coat of arms reflected both his national and his scientific heritage: images of a New Zealand kiwi bird and a Maori warrior, along with a motto borrowed from Lucretius, "*Primordia Quaerare Rerum* (To seek the first principles of things)." Fittingly, his ashes were interred in a grave at Westminster Abbey next to the final resting places of Newton and Lord Kelvin.

5

A Compelling Quartet

The Four Fundamental Forces

> The grand aim of all science . . . is to cover the greatest number of empirical facts by logical deduction from the smallest possible number of hypotheses or axioms.
>
> —ALBERT EINSTEIN (*THE PROBLEM OF SPACE, ETHER, AND THE FIELD IN PHYSICS*, 1954, TRANSLATED BY SONJA BARGMANN)

In 1939, Niels Bohr arrived at Princeton with a grave secret. He had just learned that Nazi Germany was pioneering the methods of nuclear fission: the splitting of uranium and other large nuclei. The unspoken question was whether the powerful energies of atomic cores could be used by Hitler to manufacture deadly weapons. To understand fission better, Bohr was working with John Wheeler to develop a model of how nuclei deform and fragment.

Out of respect, Bohr attended one of Einstein's lectures on unification. Einstein presented an abstract mathematical model of uniting gravitation with electromagnetism. It did not mention nuclear forces, nor did it even address quantum mechanics. Bohr reportedly left the talk in silence. His disinterest characterized the spirit of the times; the nucleus was the new frontier.

Nuclear physics had by then become intensely political. The previous year, Otto Hahn, a German chemist who had assisted Rutherford during his days at McGill, along with Lise Meitner and Fritz Strassmann had discovered how to induce the fission of a particular isotope of uranium through the bombardment of neutrons. When later that year Meitner fled the Nazis after the Anschluss (annexation of Austria), she brought word of the discovery to her nephew Otto Frisch, who was working with Bohr. Bohr became alarmed by the prospects that the Nazis could use this finding to develop a bomb—an anxiety that others in the know soon shared. These fears intensified when Szilard and Italian physicist Enrico Fermi demonstrated that neutrons produced by uranium nuclei splitting apart could trigger other nuclei to split—with the ensuing chain reaction releasing enormous quantities of energy. Szilard wrote a letter to Roosevelt warning of the danger and persuaded Einstein to sign it. Soon the Manhattan Project was born, leading to the development by the Americans of the atomic bomb.

The nucleus was a supreme puzzle. What holds it together? Why does it decay in certain ways? What causes some isotopes to disintegrate more readily than others? How come the number of neutrons in most atoms greatly exceeds the number of protons? Why does there seem to be an upper limit on the size of nuclei found in nature? Could artificial nuclei of any size be produced?

Throughout the turbulent years culminating in the Second World War, one of the foremost pioneers in helping to resolve those mysteries was Fermi. Born in Rome on September 29, 1901, young Enrico was a child prodigy with an amazing aptitude for math and physics. By age ten he was studying the nuances of geometric equations such as the formula for a circle. After the tragic death of his teenage brother, he immersed himself in books as a way of trying to cope, leading to even further acceleration in his studies. Following a meteoric path through school and university, he received a doctorate from the University of Pisa when he was only twenty-one. During the mid-1920s, he spent time in Gottingen, Leiden, and Florence, before becoming professor of physics at the University of Rome.

Among other critical contributions Fermi made to nuclear and particle physics, in 1933, he developed the first mathematical model of beta decay. The impetus to do so arose when at the Seventh Solvay Conference earlier that year, Pauli spoke formally for the first time about the theory of the neutrino. Pauli explained that when beta rays are emitted from the radioactive decay of a nucleus, an unseen, electrically neutral, lightweight particle must be produced to account for unobserved extra energy. He had originally called it a neutron, but when those heavier particles were discovered, he took up a suggestion by Fermi and switched to calling it by its Italian diminutive. Fermi proceeded to calculate how the decay process would work. Though, as it would turn out, his model was missing several key ingredients, it offered the monumental unveiling of a wholly new force in nature—the weak interaction. It is the force that causes certain types of particle transformations, producing unstable phenomena such as beta decay.

As physicist Emilio Segrè, who worked under Fermi, recalled:

> Fermi gave the first account of this theory to several of his Roman friends while we were spending the Christmas vacation of 1933 in the Alps. It was in the evening after a full day of skiing; we were all sitting on one bed in a hotel room, and I could hardly keep still in that position, bruised as I was after several falls on icy snow. Fermi was fully aware of the importance of his accomplishment and said that he thought he would be remembered for this paper, his best so far.[1]

Fermi's model of beta decay imagines it as an exchange process involving particles coming together at a point. For example, if a proton meets up with an electron, the proton can transfer its positive charge to the electron, transforming itself into a neutron and the electron into a neutrino. Alternatively, a proton can exchange its charge and become a neutron, along with a positron and a neutrino. As a third possibility, a neutron can transmute into a proton, in conjunction with an electron and an antineutrino (like a neutrino but with a different production mechanism). Each of these involves a huddling together and a transfer—like a football player approaching a member of the opposing team, grabbing the ball, and heading off in another direction.

In electromagnetism, two electric currents—streams of moving electric charge—can interact with each other by means of the exchange of a photon. Because the photon is an electrically neutral particle, no charge is transferred in the process. Rather, the photon exchange can either bring the currents together or separate them depending on the nature and direction of the moving charges.

In modern terminology, we say the photon is the exchange particle conveying the electromagnetic force. Exchange particles, including photons, belong to a class of particles called bosons. The smallest ingredients of matter—now known to be quarks and leptons—are all fermions. If fermions are like the bones and muscles of the body, bosons supply the nerve impulses that provide their dynamics.

For the weak force, as Fermi noted, two "currents," one the proton/neutron and the other the electron/neutrino, can exchange charge and identity during their process of interaction. Here Fermi generalized the concept of current to mean not just moving charges but also any stream of particles that may keep or alter certain properties during an interaction.

Just as mass measures the impact of gravity, and charge the strength of electromagnetism, Fermi identified a factor—now known as the Fermi weak coupling constant—that sets the

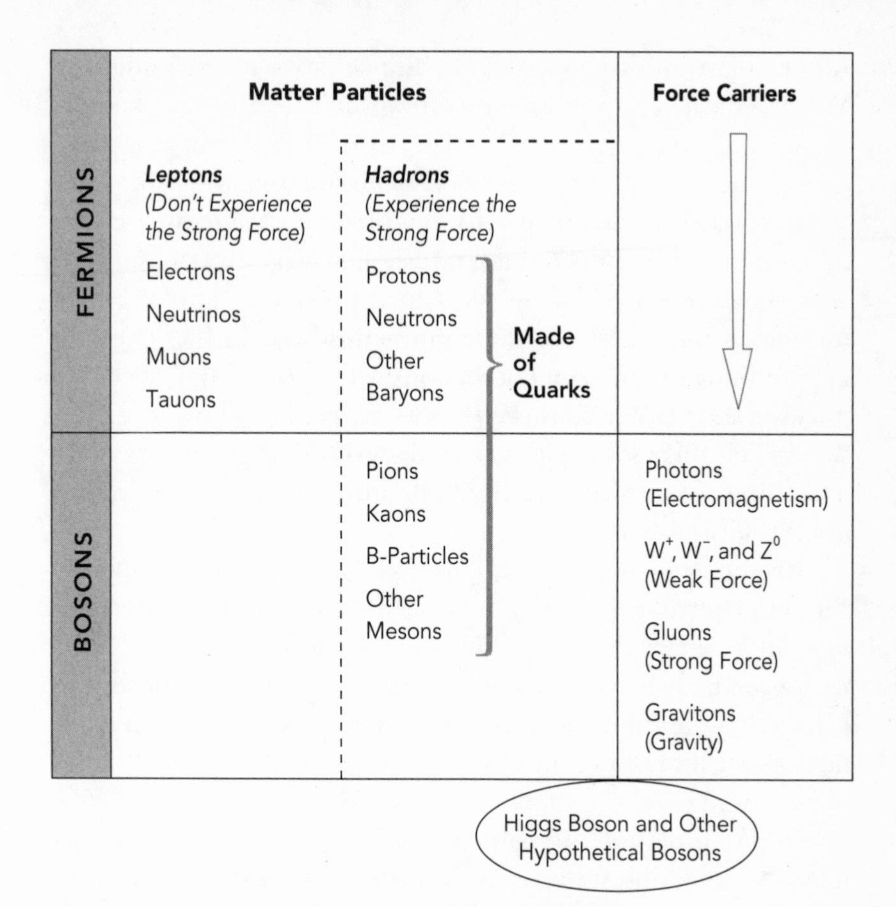

Types of elementary particles.

strength of the weak interaction. He used this information to construct a method, known as Fermi's "golden rule," for calculating the odds of a particular decay process taking place. Suddenly, the long-established gravitational and electromagnetic interactions had a brand-new neighbor. But no one knew back then how to relate the new kid on the block to the old-timers.

To make matters even more complicated, in 1934, Japanese physicist Hideki Yukawa postulated a fourth fundamental interaction, similarly on the nuclear scale. Yukawa noted that while beta decay is a rare event, another linkage between protons and neutrons is much more common and significantly more powerful.

Rather than causing decay, it enables coherence. To distinguish Yukawa's nuclear interaction from Fermi's, the former became known as the strong force.

The need for a strong force bringing together nucleons (nuclear particles) has to do with their proximity and, in the proton's case, their identical charge. Judging each other on the basis of charge alone, protons wouldn't want to stick together. Their mutually repulsive electrostatic forces would make them want to get as far away from each other as possible, like the north poles of two magnets pushing each other apart. The closer together they'd get, their shared desire to flee would grow even greater. Then how do they fit into a cramped nucleus on the order of a quadrillionth of an inch?

Born in Tokyo in 1907, Yukawa grew up at a time when the Japanese physics community was very isolated and there was very little interaction with European researchers. His father, who became a geology professor, strongly encouraged him to pursue his scientific interests. Attending the university where his father taught, Kyoto University, he demonstrated keen creativity in dealing with mathematical challenges—which would propel him to a pioneering role in establishing theoretical physics in his native land. At the age of twenty-seven, while still a Ph.D. student, he developed a brilliant way of treating nuclear interactions that became a model for describing the various natural forces.

Yukawa noted that while electromagnetic interactions can bridge vast distances, nuclear forces tend to drop off very quickly. The magnetic effects of Earth's iron core can, for example, align a compass thousands of miles away, but nuclear stickiness scarcely reaches beyond a range about one trillionth of the size of a flea. He attributed the difference in scale to a distinction in the type of boson conveying the interaction. (Remember that bosons are like the universe's nervous system, conveying all interactions.) The photon, a massless boson, serves to link electrical currents spanning an enormous range of distances.

If it were massive, however, its range would shrink down considerably, because the inverse-squared decline in interactive strength over distance represented by Maxwell's wave equations would be replaced by an exponentially steeper drop. The situation would be a bit like throwing a Frisbee back and forth across a lawn and then replacing it with a lead dumbbell. With the far heavier weight, you'd have to stand much closer to keep up the exchange.

By substituting nuclear charge for electric charge, and massive bosons, called mesons, for photons, Yukawa found that he could describe the sharp, pinpoint dynamics of the force between nucleons—demonstrating why the interaction is powerful enough to bind nuclei tightly together while being insignificant at scales larger than atomic cores. All that would be needed was a hitherto unseen particle. If Dirac's hypothesized positrons could be found, why not mesons?

Nature sometimes plays wicked tricks. In 1936, Carl Anderson observed a strange new particle in a stream of cosmic rays. Because a magnetic field diverted it less than protons and more than electrons or positrons, he estimated its mass to be somewhere in between—a little more than two hundred times the mass of the electron. On the face of things, it seemed the answer to nuclear physicists' dreams. It fit in well with Yukawa's predictions for the mass of the exchange boson for the strong force, and physicists wondered if it was the real deal.

Strangely enough, any resemblance between the cosmic intruder and Yukawa's hypothesized particle was pure coincidence. Further tests revealed the new particle to be identical to the electron in all properties except mass. Indeed it turned out to be a lepton, a category that doesn't experience the strong force at all, rather than a hadron, the term denoting strongly interacting particles. (Lepton and hadron derive from the Greek for "thin" and "thick," respectively—a reference to their relative weights that is not always accurate; some leptons are heavier than some hadrons.) Anderson's particle was eventually renamed the muon, to distinguish it from Yukawa's exchange particle. Pointing out the muon's seeming redundancy and lack of relevance to the theories of his time, physicist Isidor I. Rabi famously remarked, "Who ordered that?"

True mesons would not be found for more than a decade. Not many nuclear physicists were contemplating pure science during that interval; much energy was subsumed by the war effort. Only after the war ended could the quest for understanding the world of particles resume in earnest.

In 1947, a team of physicists led by Cecil Powell of the University of Bristol, England, discovered tracks of the first known meson, in a photographic image of cosmic ray events. Born in Tonbridge in Kent, England, in 1903, Powell had an unlucky early family life. His grandfather was a gun maker who had the misfortune of accidentally blinding someone while out shooting—an action that led to a lawsuit and financial ruin. Powell's father tried to continue in the family trade, but the advent of assembly-line production bankrupted him.

Fortunately, Powell himself decided to pursue a different career path. Receiving a scholarship to Cambridge in 1921, he consulted with Rutherford about joining the Cavendish group as a research student. Rutherford agreed and arranged for Charles Wilson to be his supervisor. Powell soon became an expert on building cloud chambers and using them for detection.

In the mid-1930s, after Cockcroft and Walton built their accelerator, Powell constructed his own and actively studied collisions between high-energy protons and neutrons. By then he had relocated to Bristol. While at first he used cloud chambers to record the paths of the by-products, he later found that a certain type of photographic emulsion (a silver bromide and iodide coating) produced superior images. Placing chemically treated plates along the paths of particle beams, he could observe disintegrations as black "stars" against a transparent background—indicating all of the offshoots of an interaction. Moreover the length of particle tracks on the plates offered a clear picture of the decay products' energies—with any missing energy indicating possible unseen marauders, such as neutrinos, that have discreetly stolen it away.

In 1945, Italian physicist Giuseppe Occhialini joined the Bristol group, inviting one of his most promising students, César Lattes, along one year later. Together with Powell they embarked upon an extraordinary study of the tracks produced by cosmic rays. To obtain their data they brought covered photographic plates up to lofty altitudes, including an observatory high up in the French Pyrenees and onboard RAF (Royal Air Force) aircraft. After exposing the plates to the steady stream of incoming celestial particles, the researchers were awestruck by the complex webs of patterns they etched—intricate family trees of subatomic births, life journeys, and deaths.

As Powell recalled:

> When [the plates] were recovered and developed at Bristol it was immediately apparent that a whole new world had been revealed. The track of a slow proton was so packed with developed grains that it appeared almost like a solid rod of silver, and the tiny volume of emulsion appeared under the microscope to be crowded with disintegrations produced by fast cosmic ray particles with much greater energies than any which could have been produced artificially at the time. It was as if, suddenly, we had broken into a walled orchard, where protected trees had flourished and all kinds of exotic fruits had ripened in great profusion.[2]

Among the patterns they saw was a curious case of one mid-size particle stopping and decaying into another, appearing as if a slightly more massive type of muon gave birth to the conventional variety. Yet a long line of prior experiments indicated that if muons decay they always produce electrons, not more muons. Consequently, the researchers concluded that the parent particle must have been something else. They named it the "pi meson," which became "pion" for short. It soon became clear that the pion matched the exchange particle predicted by Yukawa.

Around the same time, George Rochester of the University of Manchester detected in cloud chamber images a heavier type of meson, called the neutral kaon, that decays along a V-shaped track into two pions—one positive and the other negative. In short order, researchers realized that pions and kaons each have positive, negative, and neutral varieties—with neutral kaons themselves coming in two distinct types, one shorter lived than the other.

The importance of the discovery of mesons was so widely recognized that Powell received the Nobel Prize in lightning speed—in 1950, only three years later. Occhialini would share the 1979 Wolf Prize, another prestigious award, with George Uhlenbeck.

The Bristol team's discovery represented the culmination of the Cavendish era of experimental particle physics. From the 1950s until the 1970s, the vast majority of new findings would take place by means of American accelerators, particularly successors to Lawrence's cyclotron. An exciting period of experimentation would demonstrate that Powell's "orchard of particles" is full of strange fruit indeed.

While high-energy physicists, as researchers exploring experimental particle physics came to be known, tracked an ever-increasing variety of subatomic events, a number of nuclear physicists joined with astronomers in attempts to unravel how the natural elements formed. An influential paper by physicist Hans Bethe, "Energy Production in Stars," published in 1939, showed how the process of nuclear fusion, the uniting of smaller into larger nuclei, enables stars to shine. Through a cycle in which ordinary hydrogen combines into deuterium, deuterium unites with more hydrogen to produce helium-3, and finally helium-3 combines with itself to make helium-4 and two extra protons, stars generate enormous amounts of energy and radiate it into space. Bethe proposed other cycles involving higher elements such as carbon.

George Gamow, by then at George Washington University, humorously borrowed Bethe's name while applying his idea to the early universe in a famous 1948 paper with Ralph Alpher, "The Origin of Chemical Elements." Although Alpher and Gamow were the paper's true authors, they inserted Bethe's appellation to complete the trilogy of the first Greek letters; hence it is sometimes known as the "alphabetical paper."

Alpher and Gamow's theory of element production relies on the universe having originated in an extremely dense, ultrahot state, dubbed by Fred Hoyle the "Big Bang." (Hoyle, a critic of the theory, meant his appellation to be derogatory, but the name stuck.) The idea that the universe was once extremely small was first proposed by Belgian mathematician and priest Georges Lemaitre, and gained considerable clout when American astronomer Edwin Hubble discovered that distant galaxies are moving away from ours, implying that space is expanding. Alpher and Gamow hypothesized that helium, lithium, and all higher elements were forged in the furnace of the fiery nascent universe.

Curiously enough, although they were right about helium, they were wrong about the other elements. While the primordial universe was indeed hot enough to fuse helium from hydrogen, as it expanded, it markedly cooled down and could not have produced higher elements in sufficient quantities to explain their current amounts. Thus the carbon and oxygen in plants and animals were not produced in the Big Bang. Rather, as Hoyle and three of his colleagues demonstrated, elements higher than helium were wrought in a different type of cauldron—the intense infernos of stellar cores—and released into space through the stellar explosions called supernovas.

Gamow was flummoxed by the idea that there could be two distinct mechanisms for element production. In typical humorous fashion, he channeled his bafflement and disappointment into mock biblical verse: a poem titled "New Genesis."

"In the beginning," the verse begins, "God created radiation and ylem (primordial matter)." It continues by imagining God fashioning element after element simply by calling out their mass numbers in order. Unfortunately, God forgot mass number five, almost dooming the whole enterprise. Rather than starting again, He crafted an alternative solution: "And God said: 'Let there be Hoyle' . . . and told him to make heavy elements in any way he pleased."[3]

Despite its failure to explain synthesis of higher elements, the Big Bang theory has proven a monumentally successful description of the genesis of the universe. A critical confirmation of the theory came in 1965 when Arno Penzias and Robert W. Wilson pointed a horn antenna into space and discovered a constant radio hiss in all directions with a temperature of around three degrees above absolute zero (the lower limit of temperature). After learning of these results, Princeton physicist Robert Dicke demonstrated that its distribution and temperature were consistent with expectations for a hot early universe expanding and cooling down over time.

In the 1990s and 2000s, designated satellites, called the COBE (Cosmic Background Explorer) and the WMAP (Wilkinson Microwave Anisotropy Probe), mapped out the fine details of the cosmic background radiation and demonstrated that its temperature profile, though largely uniform, was pocked with slightly hotter and colder spots—signs that the early universe harbored embryonic structures that would grow up into stars, galaxies, and other astronomical formations. This colorfully illustrated profile was nicknamed "Baby Picture of the Universe."

The Baby Picture harkens back to a very special era, about three hundred thousand years after the Big Bang, in which electrons joined together with nuclei to form atoms. Before this "era of recombination," electromagnetic radiation largely bounced between charged particles in a situation akin to a pinball machine. However, once the negative electrons and positive cores settled down into neutral atoms, it was like turning off the "machine" and letting the radiation move freely. Released into space the hot radiation filled the universe—bearing subtle temperature differences reflecting slightly denser and slighter more spread out pockets of atoms. As the cosmos evolved, the radiation cooled down and the denser regions drew more and more matter. When regions accumulated the critical amount of hydrogen to fuse together, maintain steady chain reactions, and release energy in the form of light and heat, they began to shine and stars were born.

The creation of stars, planets, galaxies, and so forth is the celestial drama that engages astrophysicists and astronomers. Particle physicists are largely interested in the back story: what happened before recombination. The details of how photons, electrons, protons, neutrons, and other constituents interacted with one another in the eons before atoms, and particularly in the first moments after the Big Bang reflect the properties of the fundamental natural interactions. Therefore, like colliders, the early universe represents a kind of particle physics laboratory; any discoveries from one venue can be compared to the other.

The same year that Alpher and Gamow published their alphabet paper, three physicists, Julian Schwinger and Richard Feynman of the United States and Sin-Itiro Tomonaga of Japan, independently produced a remarkable set of works describing the quantum theory of the electromagnetic interaction. (Tomonaga developed his ideas during the Second World War when it was impossible for him to promote them.) Distilled into a comprehensive theory through the vision of Princeton physicist Freeman Dyson, quantum electrodynamics (QED), as it was called, became seen as the prototype for explaining how natural forces operate.

Of all the authors who developed QED, the one who offered the most visual representation was Feynman. He composed a remarkably clever shorthand for describing how particles communicate with one another—with rays (arrowed line segments) representing electrons and other charged particles, and squiggles denoting photons. Two electrons exchanging a photon, for example, can be depicted as rays coming closer over time, connecting up with a squiggle, and then diverging. Assigning each possible picture a certain value, and developing a means for these to be added up, Feynman showed how the probability of all manner of electromagnetic interactions could be determined. The widely used notation became known as Feynman diagrams.

Through QED came the alleviation of certain mathematical maladies afflicting the quantum theory of electrons and other charged particles. In trying to apply earlier versions of quantum field theory to electrons, theorists obtained the nonsensical answer "infinity" when performing certain calculations. In a process called renormalization, Feynman showed that the values of particular diagrams nicely canceled out, yielding finite solutions instead.

Inspired by the power of QED, in the 1950s, various theorists attempted to apply similar techniques to the weak, strong, and gravitational interactions. None of the efforts in this theoretical triathlon would come easy—with each leg of the race offering unique challenges.

By that point, Fermi's theory of beta decay had been extended to muons and become known as the Universal Fermi Interaction. Confirmation of one critical prediction of the theory came during the middle of the decade, when Frederick Reines and Clyde Cowan, scientists working at Los Alamos National Laboratory, placed a large vat of fluid near a nuclear reactor and observed the first direct indications of neutrinos. The experiment was set up to measure rare cases in which neutrinos from the reactor would interact with protons in the liquid, changing them into neutrons and positrons (antimatter electrons) in a process that is called reverse beta decay. When particles meet their antimatter counterparts, they annihilate each other in a burst of energy, producing photons. Neutrons, when absorbed by the liquid, also produce photons. Therefore Reines and Cowan realized that twin flashes (in another light-sensitive fluid) triggered by dual streams of photons would signal the existence of neutrinos. Amazingly, they found such a rare signal. Subsequent experiments they and others performed using considerably larger tanks of fluid confirmed their groundbreaking results.

By the time of the confirmation of the final component of Fermi's theory—the prototype of the weak interaction—physicists had begun to realize its significant gaps. These manifested themselves by way of comparison with the triumphs of QED. QED is a theory replete with many natural symmetries. Looking at Feynman diagrams representing its processes, many of these symmetries are apparent. For example, flip the time axis, reversing the direction of time, and you can't tell the difference from the original. Thus, processes run the same backward and forward in time. That is a symmetry called time-reversal invariance.

Another symmetry, known as parity, involves looking at the mirror image of a process. If the mirror image is the same, as in the case of QED, that is called conservation of parity. For example, the letter "O," looking the same in the mirror, has conserved parity, while the letter "Q" clearly doesn't because of its tail.

In QED, mass is also perfectly conserved—representing yet another symmetry. When electrons (or other charged particles) volley photons back and forth, the photons carry no mass whatsoever. Electrons keep their identities during electromagnetic processes and never change identities. Comparing that to beta decay, in which electrons sacrifice charge and mass and end up as neutrinos, the difference is eminently clear.

The question of symmetries in the weak interaction came to the forefront in 1956 when Chinese American physicists Tsung Dao Lee and Chen Ning (Frank) Yang proposed a brilliant solution to a mystery involving meson decay. Curiously, positively charged kaons have two different modes of decay: into either two or three pions. Because each of these final states has a different parity, physicists thought at first that the initial particles constituted two separate types. Lee and Yang demonstrated that if the weak interaction violated parity, then one type of particle could be involved with both kinds of processes. The "mirror-image" of certain weak decays could in some cases be something different. Parity violation seemed to breach common sense, but it turned out to be essential to understanding nuances of the weak interaction.

Unlike the weak interaction, the strong force does not have the issue of parity violation. Thanks to Yukawa, researchers in the 1950s had a head start in developing a quantum theory of that powerful but short-ranged force. However, because at that point experimentalists had yet to probe the structure of nucleons themselves, the Yukawa theory was incomplete.

The final ingredient in assembling a unified model of interactions would be a quantum theory of gravity. After QED was developed, physicists trying to develop an analogous theory of gravitation encountered one brick wall after another. The most pressing dilemma was that while QED describes encounters that take place over time, such as one electron being scattered by another due to a photon exchange, gravitation, according to general relativity, is a feature stemming from the curvature of a timeless four-dimensional geometry. In other words, it has the agility of a statue. Even to start thinking about quantum gravity required performing the magic trick of turning a timeless theory into an evolving theory. A major breakthrough came in 1957 when Richard Arnowitt, Stanley Deser, and Charles Misner developed a way of cutting space-time's loaf into three-dimensional slices changing over time. Their method, called ADM formalism, enabled researchers to craft a dynamic theory of gravity ripe for quantization.

Another major problem with linking gravity to the other forces involves their vast discrepancy in strength—a dilemma that has come to be known as the hierarchy problem. At the subatomic level, gravitation is 10^{40} (1 followed by 40 zeroes) times punier than electromagnetism, which itself is much less formidable than the strong force. Bringing all of these together in a single theory is a serious dilemma that has yet to be satisfactorily resolved.

Finally, yet another wrench thrown into the works involves renormalizing any gravitational counterparts to QED. To theorists' chagrin, the methods used by Schwinger, Feynman, and Tomonaga were ineffective in removing infinite terms that popped up in attempts to quantize gravity. Gravity has proven a stubborn ox indeed.

Unification is one of the loftiest goals of the human enterprise. We long for completeness, yet each discovery of commonalities seems accompanied by novel examples of diversity. Electricity and magnetism get along together just perfectly, as Maxwell showed, but the other forces each have glaring differences. The periodic table seemed fine for a while to explain the elements until scientists encountered isotopes. Rutherford, Bohr, Heisenberg, and their colleagues seemed to wrap up the world of the atom in a neat parcel, until neutrinos, antimatter, muons, and mesons arrived on the scene.

From the mid-1950s until the mid-1990s, powerful new accelerators would reveal a vastly more complex realm of particles than anyone could have imagined. Suddenly, ordinary protons, neutrons, and electrons would be vastly outnumbered by a zoo of particles with bizarre properties and a wide range of lifetimes. Only a subset of the elementary particles could even be found in atoms; most had nothing to do with them save their reactions to the fundamental forces. It would be like walking into a barn and finding the placid cows and sheep being serenaded by wild rhinoceri, hyenas, platypi, mammoths, and a host of unidentified alien creatures. Given the ridiculously diverse menagerie nature had revealed, finding any semblance of unity would require extraordinary pattern-recognition skills, a keen imagination, and a hearty sense of humor.

6

A Tale of Two Rings

The Tevatron and the Super Proton Synchrotron

It is true that there were a few flaws in my logic. The rivers of ground water that flowed through their experiments, the walls of piling rusting away, the impossible access, and all without benefit of toilet facilities. But some of the users had their finest moments down in these proton pits—the discovery of beauty, the bottom quark, where else?! Alas, as far as I know not one piling has been pulled up, not one pit has yet been refilled with earth.

—ROBERT R. WILSON (LECTURE AT THE THIRD INTERNATIONAL SYMPOSIUM ON THE HISTORY OF PARTICLE PHYSICS, STANFORD LINEAR ACCELERATOR CENTER, JUNE 1992)

After finding the misplaced rubber gasket that had stopped up his cyclotron during an important demonstration of its medical uses, Lawrence was absolutely furious. "You get out of this laboratory!" he screamed at the young assistant. "Don't you ever come back!"[1]

Robert R. (Bob) Wilson, a graduate student at the Berkeley Radiation Laboratory who would blossom into the designer and leader of the greatest enterprise in the history of American high-energy physics, was absolutely crushed. With the laboratory dressed up in hospital white, and patients ready to be treated for their cancerous tumors, how could he have been so careless? Patients were literally waiting for days while the cyclotron refused to work, all because of his silly mistake. No words could express the depth of his remorse.

Lawrence rehired Wilson, only to fire him again after he ruined a pair of expensive pliers by melting them in a hot flame. The second dismissal wasn't quite so bad. "I thought I'd probably get back somehow," Wilson recalled.[2]

To say that Wilson's career took many twists and turns before he became the force behind the establishment of Fermilab, the foremost accelerator lab in the United States and for a time in the world, is an understatement. Born in Frontier, Wyoming, in 1914, he came, like Rutherford and Lawrence, from a pioneering family. Wilson's mother, Edith, was the daughter of a rancher who had come to the region during a gold rush. She married Wilson's father when he was working in town as a surveyor. A stern, practical man, he could never appreciate the academic proclivities of his son. When Wilson, who was a well-read youth, wanted to head off to Berkeley for college, his father tried to forbid him—insisting that he go into business instead. Unsupported by his dad, Wilson set out at the age of eighteen for a life of adventure in particle physics.

At Berkeley, eyeing Lawrence's lab for the first time, Wilson marveled like a child at a Christmas display. He was awestruck by the fancy equipment and the researchers' exuberance. Although on a personal level he found Lawrence egotistical, at least initially, he decided to pursue working at the Rad Lab for his undergraduate research project. Nervously, he traipsed over to Lawrence's office to ask about a position, and was greatly relieved when the great director replied, "Oh, yes, yes, yes."

Wilson became an expert at cyclotron design, particularly with regard to producing stable particle orbits. He completed his undergraduate studies at Berkeley and continued as Lawrence's graduate student. He witnessed the Rad Lab becoming a shining example for high-energy research around the world—its physicists respected for their experience.

Wilson also learned a great deal from Lawrence's leadership skills. "I'm sure he had a profound influence on me," Wilson recalled. "His style of running that laboratory was very impressive. He led by example. . . . We were infected by his enthusiasm and optimism, and his sense of priorities and of pushing hard."[3]

In 1940, after getting his Ph.D. from Berkeley, Wilson got married to California native Jane Scheyer and moved east with her to Princeton, New Jersey. He served there as instructor for three years before being recruited to Los Alamos to work on the Manhattan Project. After the war, he spent a year at Harvard, then became appointed director of the Laboratory for Nuclear Studies at Cornell, where he served from 1947 to 1967. There he ran four different electron synchrotrons—the final one yielding energies up to 12 GeV—and established a reputation as an outstanding supervisor.

Synchrotrons, invented in the 1940s, offer far more flexibility than conventional cyclotrons by stepping up their magnetic fields in tandem with the requirements of the packs of particles rounding their tracks. Particles are injected into a synchrotron in bunches, like cyclists clustered together during a race. As each grouping reaches higher energies, the magnetic field is ramped up so that instead of spiraling outward the particles maintain the same radial distance. Fiercer creatures require stronger leashes.

Another key characteristic of a synchrotron is that its central magnet is replaced by bending magnets placed at equal intervals around the beam path. These serve a similar purpose but enable the device to encompass a wide area (a football field or farmlands, for example), thus permitting a much larger radius and increasing its power well beyond room-size instruments.

Yet another difference between synchrotrons and the original cyclotrons has to do with variations in the driving electric field. While the electric fields of cyclotrons vary periodically, offering a constant cadence of boosts, those of synchrotrons keep pace with the particle packets—preventing them from falling out of step once they reach relativistic speeds. It's like a father who is pushing his son on a swing increasing his rhythm for greater effect after it has sped up. Similarly, synchrotrons are flexible enough to raise the energy bar of already energetic particles.

Wilson's tenure at Cornell coincided with a dramatic rise of the use of synchrotrons in particle physics. The extraordinary power and flexibility of synchrotrons would prove critical for the discovery of massive new types of particles. Large synchrotrons would ultimately supply the dynamos for mighty particle colliders that would be used to search for evidence of unity—such as identifying the exchange particles for electroweak unification. Over decades, researchers at various laboratories would find ways of increasing synchrotrons' ring sizes and improving the focusing power of their magnets to make them more effective at producing high-energy particles.

The 1950s and 1960s were a golden age for synchrotron design. In Berkeley, two engineers under Lawrence's supervision, William Brobeck and Edward Lofgren, constructed a concert-hall-size proton synchrotron called the Bevatron. Completed in 1954, it could reach energies of up to 6 GeV. Unfortunately, its costs were driven up by a doughnut-shaped vacuum chamber (and surrounding magnet) with such wide openings, it seemed made for race cars rather than for particles.

Another early synchrotron, the Cosmotron, built on a converted army base in bucolic Brookhaven, New York, was more efficiently designed—possessing apertures that, though narrow, had sufficient room to accommodate the particle beams. Team leaders M. Stanley Livingston, Ernest Courant, John Blewett, G. Kenneth Green, and others managed to perform this feat through the use of 288 C-shaped magnets that carefully guided the proton pulses through the pipe of the seventy-five-foot diameter accelerator. It took but a second for the protons to travel 135,000 miles (through millions of revolutions) and reach energies of 3 GeV before smashing into targets.[4] When the Cosmotron first came on line in May 1952, the *New York Times* applauded its inaugural "Billion Volt Shot."[5]

Courant's experience with adjusting the magnets of the Cosmotron to focus the beam as tightly as possible led him to a critical insight that paved the way for the next generation of machines. He calculated that by switching adjacent magnets to face in opposite directions—alternatively inward and outward—he could greatly augment their focusing power. His finding, called strong focusing, paved the way for the construction at Brookhaven of the Alternating Gradient Synchrotron, an even mightier accelerator that opened in 1960 and is still in use today.

Meanwhile, at the European Organization for Nuclear Research (CERN) in Geneva, Switzerland, Niels Bohr smashed a bottle of champagne and inaugurated the Proton Synchrotron (PS), another strong-focusing accelerator. It was a triumph for the reemergence of European science after the war. CERN had been established a decade earlier, through a resolution put forth by Rabi at the fifth UNESCO conference authorizing that agency, "to assist and encourage the formation and organization of regional centres and laboratories in order to increase and make more fruitful the international collaboration of scientists."[6]

By the time the PS opened, the CERN council, consisting of representatives of Belgium, Denmark, France, (West) Germany, Greece, Italy, the Netherlands, Norway, Spain, Sweden, Switzerland, the United Kingdom, and Yugoslavia, had met numerous times, and established its scientific laboratory near the village of Meyrin—part of the Geneva canton, close to the French border. The smashing new accelerator bolstered the center's reputation as an international hub of high-energy research.

Wilson worked hard at Cornell to compete with the burgeoning laboratories at Berkeley, Brookhaven, and CERN. He established a reputation as a highly capable leader. Yet he wanted to be more than just a scientist and administrator. In a highly unusual move for a scientist of such ambition, Wilson took time off to pursue a second career as a sculptor. In 1961, he traveled to Rome and enrolled at the Accademia di Belle Arti, where he learned how to create modern sculpture. He was also interested in architecture and other aspects of design.

All of Wilson's passions beautifully merged together when, in 1967, he was offered the supreme challenge of designing the foremost accelerator laboratory in the United States, to be built among the cornfields of rural Batavia, Illinois—about thirty-five miles west of Chicago. Originally called the National Accelerator Lab, it was renamed in 1974 after Fermi. In taking on the responsibility, Wilson set out to make the lab as user-friendly as possible—open to whoever wanted to conduct experiments requiring high energies, without regard for hierarchy. He wanted to avoid the restrictions of having just a handful of leaders, in the mode of Rutherford and Lawrence, setting the course for all of the lab's projects.

A second goal of Wilson's, meshing well with the democratic spirit he established, was to keep construction and operating costs as low as possible. The U.S. Atomic Energy Commission requested that construction time be kept to less than seven years, with a maximum expense of $250 million (reduced, due to cost-cutting measures, from an original allotment of $340 million). Miraculously, Wilson finished ahead of schedule and within the tight budget—all while doubling the accelerator's energy from an anticipated 200 GeV to more than 400 GeV. He truly wanted the greatest bang for the buck.

Despite financial restraints, Wilson fervently aspired to bring aesthetics to lab design. He involved himself in all aspects of planning the lab's architecture—including a futuristic central tower of concrete and glass—and even crafted innovative sculpture to beautify the landscape. A painter was recruited to bathe the equipment in bright colors. Unprecedented for an experimental facility, its art won accolades from the *New Republic*'s critic Kenneth Everett, who called it a "rare combination of artistic and scientific aspiration."[7] With tastes reflecting the utilitarian spirit of the 1960s, Wilson's creations managed to impress while not sapping budgets.

Finally, as a born frontiersman, he believed in living in harmony with the land and was an early environmentalist. He recycled many of the original barns by converting them to buildings used for social functions, housing, and other purposes. Wildlife from mallard ducks to muskrats found refuge in the ponds, the fields, and even the machinery. As a reminder of his roots, and also as a symbol of the pioneering nature of modern physics, he brought in a herd of bison to graze freely in a grove. Their shaggy descendants still roam the grounds today.

Wilson felt very much at home wandering Fermilab on horseback—as if it were a ranch that happened to be raising protons and mesons instead of cattle and sheep. Dressed in jeans, a windbreaker, cowboy boots, and a black fedora, he'd mount his gray mare, Star, and ride her around his enterprise—as if warming up for the Preakness—to inspect its fine details.

An aerial view of Fermi National Accelerator Laboratory (Fermilab) showing the main ring and principal office tower.

No aspect of the accelerator center was too trivial for Wilson to tweak—from the distinctive geometric rooftops (a geodesic dome in one case) to the no-frills dirt floors—even the specifics of how the kitchen was run. On a limited budget choices needed to be made—roofs versus floors, for instance—and Wilson thought that he needed to make these himself, lest he incur the wrath of the Atomic Energy Commission.

As one of the bulwarks against excess spending, Wilson hired a hardheaded administrative assistant named Priscilla Duffield, who was formerly Lawrence's secretary at the Rad Lab and then J. Robert Oppenheimer's secretary at Los Alamos during the Manhattan Project. J. David Jackson, acting head of theoretical physics at Fermilab from 1972 to 1973, remembered her as a "tall, imposing, no-nonsense woman." She would become incensed by the mere hint of unauthorized expenditure. Jackson recalled her sharp reaction when she found out about a wine-and-cheese seminar he helped organize.

> She stormed into my office, looking for my scalp. "What do you think you're doing, serving wine at that seminar? Don't you know it's illegal to spend government money on such things?" I said that I wasn't spending government money on the wine. She said, "Well who *is* paying for it?" I said, "I am." And she said, "Oh." It was the one time I saw Priscilla just a little bit penitent.[8]

Not every decision Wilson made worked out for the best, due to his passionate effort to cut costs. By steaming single-mindedly ahead with certain structural choices, he almost launched the whole project over a precipice. Not considering that magnets might sometimes need to be replaced, he had them welded to the beam line running through the tunnel of the main ring.[9] Neglecting to protect the tunnel from the humid Illinois summers, some of the magnets acquired moisture and started to crack. Imagine the horror of the eager researchers primed for discovery when right before the accelerator was about to be turned on many of its magnets failed and couldn't easily be removed. Fortunately, a stalwart team of experimentalists gathered quickly and resolved the problem.

In contrast to the permanently secured magnets, Wilson made the opposite choice for the places where physicists would take measurements. To ensure minimal cost and maximal flexibility, he designed the working areas to be as provisional as anthills. He came to realize that his makeshift structures, though thrifty, were not very popular. As Wilson remarked:

> These enclosures are indeed rough-and ready places. . . . Indeed some of the users were advised by their older colleagues to "abandon all hope, ye who enter here!" I fear that I bear the responsibility for this fiasco. In the frenzy of saving big bucks, I had the fantasy of not putting up (or down) any laboratory building at all. Instead the idea was that, once an experiment had been accepted, an outline of the necessary space would be drawn in an empty field at the end of one of the proton beams, then steel interlocking piles would be driven . . . down to the necessary depth. . . . The experimental equipment would be lowered to a luxurious graveled floor, and finally a removable steel roof would be covered with the requisite thickness of earth. . . . Simple and inexpensive, is it not? I still find it difficult to understand why all those users stopped speaking to me.[10]

Wilson nicknamed the experimental area for proton studies, constituting the termini of beams shunted from the four-mile-long main synchrotron ring, the "proton pits"; other research zones were dedicated to mesons and neutrinos. He was especially proud of the fifteen-foot bubble chamber, lauded by Berkeley physicist Paul Hernandez as the "Jewel in the Crown" for its type of detector.[11]

A bubble chamber consists of a large vat of liquid hydrogen surrounded by an immense guiding magnet. After protons collide, the magnet would steer charged debris along swirling paths

The Big European Bubble Chamber, a device for tracking particles, on display at CERN's Microcosm museum.

through the fluid. The hydrogen along the tracks would bubble away, enabling experimentalists to photograph these trails and calculate the properties of the particles that produced them. Because their different charges would experience opposite magnetic forces, positive and negative particles would spiral in opposite directions.

Other types of detectors commonly used in high-energy physics include scintillation counters, photomultipliers, Cherenkov detectors, calorimeters, spark chambers, and drift chambers. The reason for such a diverse toolbox of measuring instruments is to glean as much information as quickly as possible. Many particles, once born, are extremely short lived, and decay almost immediately into other particles. Sometimes the only signs of an interaction are missing energy, momentum, and other conserved quantities. Like detectives at a crime scene, physicists investigating possible culprits must cordon off the collision site—by surrounding as much of it as possible with data-collecting instruments—and rapidly gather a cache of evidence. Only then can they hope to reconstruct the event and determine what actually transpired.

Scintillation counters, an update on Rutherford's favorite technique for spotting particles by means of flashes in fluorescent materials, rely on atomic electrons becoming energized by passing particles and then releasing this energy as light. Moldable, fluorescent plastics and liquid additives called fluors have served well for this purpose. Photomultipliers are electronic devices that amplify faint light (from scintillators, for example) so it is much more discernable.

Cherenkov detectors depend on a different physical property, called the Cherenkov effect. Discovered in 1934 by physicist Pavel Cherenkov, of the Lebedev Physical Institute in Moscow, it is a phenomenon that occurs when a particle travels faster than light does in a particular material. Although nothing can exceed the speed of light in a vacuum, light moves slower in certain substances and can thereby be outpaced. Like jets creating a sonic boom (an audible shock wave) when they exceed sound's velocity in air, particles racing past light's material speed emit a cone of radiant energy in the direction of motion called Cherenkov radiation. Conveniently, the angle of the cone directly depends on the particle's velocity, offering a practical way of measuring this important factor.

Another class of apparatus, called calorimeters, enables researchers to record the energies of particles. These are dense materials that trigger a cascade of decays, through processes such as pair production (creation of electron-positron companions) and bremstrahlung (radiation generated when particles slow down), releasing a storehouse of energy in the process. By trapping some or all of this energy, physicists can try to determine how energetic the original must have been. While electromagnetic calorimeters rely on electromagnetic processes to produce the cascades, hadronic calorimeters depend on the strong interaction instead.

Hadrons are particles that experience the strong force, such as protons, neutrons, various types of mesons, and an assortment of heavier particles. They are each composed of quarks. Leptons such as electrons, positrons, muons, and neutrinos, on the other hand, are particles that ignore the strong force. They are not made of quarks but, rather, are fundamental. Hadronic calorimeters capture the energy of hadrons but not leptons.

Along with bubble chambers, various other types of instruments can be used to measure particle trajectories. Spark chambers, useful for charged particles, involve electrical signals passing like lightning through regions of a gas ionized by particles whizzing past. Drift chambers are more sophisticated devices that use electronics to record the time particles take to move from one point to another.

The invention of the computer provided a vital tool for high-energy research. It allowed researchers the luxury of sifting through vast amounts of data and accessing which subset displayed the fingerprints of potentially interesting events. Otherwise finding rare decay products would be a hopeless task—like locating a single four-leaf clover in the vast expanse of the American prairie.

By the time Fermilab became operational in the early 1970s, one of its principal features was already out of date. Along the lines of Rutherford's experiments, beams produced by the accelerator slammed into fixed targets. According to conservation principles, most of the collision energy served to channel secondary particles along a tight path past the target. Merely a fraction of the energy could be used to produce the new particles themselves. Technically, this is because the useful energy for a fixed-target collision increases at the relatively slow rate of the square root of the beam energy. If, with improvements to a fixed-target device, protons were accelerated to one hundred times more energy, for instance, its effective energy would increase only tenfold. Not only was this situation inefficient, the narrowness of the particle jets produced made it difficult for researchers to examine what was created.

As far back as 1953, Wideröe, with his extraordinary foresight, had patented a design for a far more efficient type of accelerator, now known as a collider.[12] He recognized that by smashing particles together head-on, a much larger portion of the collision energy would be transformative rather than kinetic (engendering motion). Because he was working as an industrial engineer at the time, the physics community did not take note of his patent. Three years later, however, a team of synchrotron developers, led by physicist Donald Kerst, independently proposed the idea of colliding particle beams with each other. Published in the leading journal *Physical Review* and discussed at a 1956 CERN symposium in Geneva, Kerst's proposal stimulated efforts to boost the usefulness of conventional accelerators by turning them into colliders.

We can envision the difference between fixed-target accelerators and colliders by imagining two different kinds of accidents involving diesel locomotives. In the first case, picture an engine car racing out of control and hitting the back of a boxcar that's sitting at the junction of two tracks. Conceivably, the impact of the engine car could cause the boxcar to roll down one of the tracks, the engine car down the other, and both would escape unscathed. The collision energy would be mainly kinetic.

However, suppose two engine cars (of comparable size and speed), traveling in opposite directions, plow into each other head on. In that case it would be hard to picture a happy outcome. The bulk of the energy would likely end up producing a burning wreck. What would be horrific for a transportation engineer would work out nicely for high-energy physicists in their quest to add more fuel to the fire and spark the creation of new particles.

At the same CERN conference, Princeton physicist Gerald O'Neill proposed a clever way of implementing the collider idea by use of linked storage rings. Particles accelerated in a synchrotron, he envisioned, could be directed into two different storage rings, where they would orbit in opposite directions before smashing together at a designated intersection point. His idea formed the basis of several important electron-positron collision projects during the 1960s and early 1970s, culminating in the completion of the SPEAR ring at SLAC in 1972—where Burton Richter, SPEAR's developer, would codiscover the J/psi particle (a heavy meson composed of quarks with properties called "charm" and "anticharm"), and Martin Perl would discover the ultraheavy tau lepton, among other findings.

These discoveries would help provide evidence that quarks and leptons are organized into three distinct generations: the up and down quarks, electron and neutrino in the first; the strange and charm quarks, muon and muon neutrino in the second; and the tau lepton (along with the later-discovered top and bottom quarks and tau neutrino) in the third. Either individually, as in the case of leptons, or grouped into various combinations to form different hadrons, as in the case of quarks, these constitute the basis of matter.

In 1971, CERN inaugurated the world's first hadron collider: the Intersecting Storage Rings (ISR). CERN's then existing Proton Synchrotron accelerated bunches of protons to energies of 28 GeV, upon which an injection system whisked them off to one of two storage carousels. There, they were "stacked," a process involving timing the proton injections so that groups are packed in closely together but still flowing smoothly. It's a bit like a traffic signal allowing cars to merge onto a highway only at particular intervals to pace them just right and increase the road's capacity. Through stacking, the proton beams circulating around the rings increase their luminosity, or rate of collisions per area, a function of the beam intensity. Boosting the luminosity is akin to upping the firing rate and focus of a machine gun to maximize its chances of hitting a target. A higher collision rate increases the chances of exceptional events taking place, such as the production of rare particles.

Shortly after the ISR came online, CERN researchers decided to test out a novel method for increasing luminosity, called stochastic cooling. Developed by Dutch physicist Simon van der Meer, who was in charge of the steering magnets at CERN, it offered a way to tighten up the bunches of protons into denser clusters, allowing them to be stacked much closer together. The basic idea is to test how far particles deviate from the average of their group and kick them back in line if they stray too far. These correcting nudges cause each bunch to have less prominent fluctuations and "cool down" to a more tightly packed state—creating more room to stack more clusters and increase the beam intensity. Van der Meer's enhancement of beam luminosity represented such an important enhancement for colliders—opening the door to pivotal discoveries—that it would earn him the 1984 Nobel Prize in Physics (along with Italian physicist Carlo Rubbia).

Anticipating the competition CERN would provide with its radically improved methods, Wilson argued for upgrades to the Fermilab accelerator that would at least double its effective energy. The reason was clear. A critical advance in theoretical physics, the unification of electromagnetism and the weak interactions into a single quantum theory, had triggered an intense race to discover the massive particles it predicted. The chance to verify a stunning new form of unity inspired a whole generation of experimentalists to join teams at Fermilab, CERN, and elsewhere and dedicate themselves to an epic search through unprecedented quantities of data generated through extraordinary energies.

The Standard Model of electroweak unification, proposed independently in 1967 by Steven Weinberg and Abdus Salam, predicts four new massive bosons, to supplement the familiar massless photon. Two of these, the W^+ and W^-, serve as the exchange bosons for the weak interaction involving positive and negative charge transfers respectively (for example, interactions involving electrons and neutrinos, or positrons and antineutrinos). A third, the Z^0, conveys the neutral version of the weak interaction. This was inserted, based upon Sheldon Glashow's work, to make it a mathematically balanced theory, even though no one at the time had ever observed a neutral weak current. Together, the W^+, W^-, and Z^0 are known as the intermediate vector bosons, the designation "vector" referring to their particular transformative properties. The fourth predicted particle is the Higgs boson, which through its spontaneous symmetry breaking (as discussed in chapter 2), supplies mass to the W^+, W^-, and Z^0 bosons, along with the quarks and leptons.

The scenario sketched by Weinberg and Salam meant that finding these new bosons wouldn't be easy. At high-enough temperatures—during the initial instants of the Big Bang, for example—the theory's symmetry would be unbroken and the W and Z bosons would be massless too. However, below a critical temperature—today's conditions, for example—the spontaneous breaking of the original symmetry would give ample mass to these bosons. To detect them, therefore, would require the extraordinarily energetic conditions of the world's mightiest accelerators.

In 1970, three intrepid experimentalists—Carlo Rubbia, then at Harvard, Alfred K. Mann, of the University of Pennsylvania, and David Cline, of the University of Wisconsin—initiated a fledgling effort at Fermilab to find the W boson. Nicknamed the HPWF collaboration, after the initials of the universities involved (and Fermilab), the group set up shop in the neutrino building. The geodesic roof of that metal structure leaked badly during rainstorms and the floors were dirt, so team members often needed to wade through muddy puddles to get to their equipment. Wilson's cost-cutting measures had led to working conditions suitable for one of Dante's lower circles.

The following year, a young virtuoso in field theory, Gerard 't Hooft of the University of Utrecht, Holland, working under the supervision of Martinus Veltman, proved that the Weinberg-Salam theory could be renormalized (infinite terms canceled out), just like quantum electrodynamics. This made the theory extremely attractive. Giddy from these remarkable results, Weinberg was eager to have one of the basic predictions of electroweak theory tested: the existence of neutral weak currents. As Rubbia later recounted, Weinberg "brainwashed" the HPWF team to switch course and look for neutral currents instead.[13]

Rubbia asked Larry Sulak, a colleague from Harvard working with the group, to install a new trigger for the detector that would be sensitive to neutral current events. These would involve fermions keeping their own identities as they interact with each other through the weak force—for example, electrons remaining electrons and protons remaining protons. The problem was that common electromagnetic interactions similarly preserve particle characteristics; electrons stay electrons during those events too. Therefore, the major challenge was to find the weak neutral needle among the haystack of electromagnetic events that similarly conserve charge and mass. Neutrino events offered the best chance for this, because as light neutral leptons their principal mode of interaction is the weak force. If a neutral hadron, such as a neutron, interacted with a neutrino in an event that kept both particles the same, the weak neutral current would be the natural culprit.

A competing team from CERN, led by Jack Fry and Dieter Haidt, also took up the gauntlet. Using the Gargamelle heavy-liquid bubble chamber, pumped full of freon, that had recently been installed at CERN's Proton Synchrotron inside a colossal superconducting magnet, they spent the fall and winter of 1972 searching for neutrino-induced neutrons. As skiers etched zigzag tracks in the snowy slopes near Geneva, Fry and Haidt examined the frozen trails of cascading particles—their specific paths in the bubble chamber marking their interactions and properties. The summer brought new joys. By July 1973, the group had collected sufficient evidence of neutral current events to present its work. The HPWF collaboration announced its own promising results around the same time.

Alas, summer's heat sometimes shapes cruel mirages. After modifying its equipment and retesting its data, the HPWF team's findings vanished amid the desert sands of statistical insignificance. Skeptics wondered if electroweak unity was simply a beautiful illusion.

Nervous that its own results would become similarly wiped out, the Gargamelle group set out for more testing, and found, to its delight, that its findings were on firm footing. Meanwhile, the HPWF group reexamined its results one more time, resolved the issues that had plagued its analysis, and proclaimed success as well. For the first time in the history of science, a wholly new mode of interaction—the neutral weak current—had been anticipated by theory before being found by experiment.

Haidt later described the impact of his team's findings: "The discovery of weak neutral currents . . . brought CERN a leading role in the field. The new effect marked the experimental beginning of the Standard Model of electroweak interactions, and triggered huge activity at CERN and all over the world, both on the experimental and theoretical sides."[14]

From the neutral current results, theorists developed fresh estimates of the mass of the W boson, stimulating an international race to find that particle. Leading the pack was an energized CERN, anxious to prove that its neutral current victory was no fluke. The European community had already identified the land and allocated the funds to begin constructing the Super Proton Synchrotron (SPS), a four-mile-long accelerator intended to be—at 300 GeV—the most energetic in the world. In the midst of construction, however, Fermilab's Main Ring surpassed the SPS's intended energy—a major disappointment for the Europeans.

When you are locked in battle, even the smallest delay can offer the other side an opening for victory. In the case of the race to identify the weak bosons, CERN's opportunity arose when Rubbia became exasperated by the failure of Fermilab to commit to building a proton-antiproton collider—an idea initially suggested by his young Harvard colleague Peter McIntyre and then developed in a 1976 paper by Cline, McIntyre, and himself. The three of them urged Wilson and the Fermilab program committee to plan out a means for running proton and antiproton beams through the same ring in opposite directions. Through enough high-energy smash-ups, they proposed, perhaps somewhere in the debris would lie the sought-after particles.

At that time, Wilson was committed instead to building the Tevatron—the world's first synchrotron with superconducting magnets guiding the beams. The name Tevatron derives from its goal of energizing protons up to 1 TeV (one teravolt or one trillion electron volts). The Tevatron would indeed be used as a collider, but until the superconducting technology was tested, the fiscally conservative director didn't want to offer a firm guarantee.

Like a persistent salesman, Rubbia knocked on CERN's door next. Because he had worked there in the 1960s, he was very familiar with the organization. Born in Gorizia, Italy, in 1934, he had come to CERN for the first time when he was twenty-six, following a university education in Pisa and Milan and a year and a half at Columbia in the United States. Since the early 1970s, Rubbia had kept up an impossibly frantic schedule, spending time at Harvard, Fermilab, and CERN. These formative experiences, coupled with a natural self-assurance, offered him the clout to suggest CERN's next move.

In shaping a new direction for CERN, Rubbia found the perfect partner in Simon van der Meer. Rubbia realized that the Dutch physicist's stochastic cooling technique would offer an ideal means of fashioning dense proton and antiproton beams. This would enable the two beams to circle in contrary ways through the SPS—greatly augmenting its center-of-mass energy by transforming it into a collider. Persuaded by Rubbia's cogent reasoning, Leon van Hove, CERN's codirector at the time, helped shepherd the concept through the bureaucracy. With amazing speed, researchers assembled the Antiproton Accumulator, a revolutionary means of building up an intense beam of those particles, and linked it to the SPS. By 1981, only five years after Cline, McIntyre, and Rubbia had proposed the idea, the SPS became operational as a proton-antiproton collider—the major purpose for which it was used throughout the decade. Although the more powerful Tevatron opened in 1983, its collider operations wouldn't begin until 1985, offering CERN a significant head start.

With the SPS and later the Tevatron came the rise of "supergroups" of researchers—teams representing dozens of institutions and hundreds of experimentalists each. High-energy physicists gravitated to the two centers hoping to share the glory of wrapping up the Standard Model, completing the third generation of fundamental particles (supplementing the tau lepton and the bottom quark—the latter found by Leon Lederman and his colleagues at Fermilab in 1977), and perhaps even discovering wholly unexpected new ones. Gone were the days of Rutherford and Lawrence—where experimental papers included but a few authors—perhaps the supervising professor, a postdoctoral researcher, and a couple of graduate students. With the burgeoning teams, some articles even included the long list of authors as lengthy footnotes. You practically needed a magnifying glass to see which contributors lent their expertise to which projects.

Moreover, because of the increasing specialization associated with the new particle-production "factories," and the many years often required to obtain results, research supervisors began to adopt a more flexible attitude toward what constituted acceptable Ph.D. theses. For example, completing Monte Carlo simulations (using a random-number generator to predict possible outcomes), writing software, building and testing new detectors, and so forth, could constitute elements of approved dissertations. Otherwise, not only wouldn't there have been enough theses to accommodate all of the graduate students working in each experimental supergroup, but also the time to get such degrees while waiting for final data would have, in many cases, been inordinately long.

Upon the reinauguration of the SPS as a collider, two detectors were readied for service—each supported by its own extensive team. The first, UA1 (Underground Area 1), was Rubbia's brainchild—an extraordinarily complex instrument that made use of state-of-the-art electronics to probe collisions from almost every possible angle. The property of covering almost the entire solid angle, known as "hermeticity," became a mainstay of detectors from that point on. No one had seen such a massive detector before—at approximately two thousand tons it was truly a Goliath. Its complexity and bulk girth inspired a French newspaper to root for the second, smaller detector UA2 (Underground Area 2), as the dexterous "David" that would slay the unwieldy giant. Reportedly, this characterization infuriated Rubbia, who considered himself the true maverick.[15]

The initial run, in December 1981, was dedicated to testing some of the predictions of quantum chromodynamics (QCD), the leading gauge theory of the strong interaction. Developed in the 1970s, QCD models the interactions among the quarks in hadrons, mediated via exchange particles called gluons. Through volleying gluons among one another, quarks of different colors cement their connections—forming baryons or mesons. The gluon concept replaced the idea of pion exchange, which failed to explain why quarks like to assemble in certain groupings and are never found roaming freely. The UA1 and UA2 detectors looked for the hard knocks of quark kernels against one another, measured the energy produced, and compared the results to theoretical QCD models. Splendidly, particularly in the UA2 results, many of the QCD predictions proved right on the mark.

After savoring the delectable appetizer of the QCD findings, it would be time for the main course. The W and Z bosons were ripe for the plucking and—thanks to the capabilities of the upgrading SPS—it would finally not be a stretch to reach for such exotic fruit. The sensitive detectors of each group were primed to taste the characteristic flavor combinations of the rare morsels.

In the case of the W bosons, the researchers expected that the quarks and antiquarks from the protons and antiprotons (up and antidown, for example) would unite briefly at high energies to produce these exchange particles. These would be extremely short lived, almost immediately decaying into charged leptons and neutrinos. Particles too fleeting to be directly detected are called resonances—manifesting themselves only through peaks in production at particular energies corresponding to their masses. It's like trying to find signs of a snowman built (from freezer scrapings) during a hot summer day; a sufficiently sized puddle of water would be a giveaway.

Around Christmas 1982, the SPS was colliding protons and antiprotons with astonishing beam luminosities of more than 10^{29} (1 followed by 29 zeroes) incident particles per square inch each second. In the UA1 detector, these yielded about 1 million events interesting enough to trigger data collection. Of these, six events met the criteria (particular amounts of energy and momentum associated with electrons fleeing at certain angles) to represent W boson candidates. Further data narrowed down the mass of the W boson to be approximately 81 GeV/c^2 (divided by the speed of light squared, in accordance with Einstein's famous mass-energy equation). Meanwhile, UA2 gathered four candidate events, confirming the important discovery.

Snagging the Z boson happened just a few months later, during a run in April/May 1983. This time, the teams looked for a different signal: the production of electron-positron pairs of particular energies. UA1 found the mass of the Z to be approximately 95.5 GeV/c^2—with the UA2 group corroborating this result. Papers in *Physics Letters B* triumphantly announced these findings, much to the delight of the physics community around the world. The discoveries were so telling, no one from that point on could question the reality of electroweak unity.

As CERN researcher Daniel Denegri, a member of the UA1 collaboration, recalled the exhilaration of the day: "This period, around the end of 1982 and throughout 1983, was an amazing time from both a professional and personal point of view. It was an unforgettable time of extreme effort, tension, excitement, satisfaction and joy."

The boost to European morale because of the weak boson findings cannot be overestimated. After decades of looking to the United States as the main innovator in high-energy physics, the continent of Einstein, Bohr, and the Curies finally got its groove back. As Denegri noted, "The discovery of the W and Z at CERN . . . signaled that the 'old side' of the Atlantic regained its eminence in particle physics."[16]

Although American researchers were happy for their colleagues across the ocean, and pleased that electroweak unification held up under close scrutiny, they could not conceal their disappointment that CERN had beaten them to the punch. Like baseball, accelerator physics had become an American pastime, so it was like losing the World Series to Switzerland. A *New York Times* editorial laid out the score: "Europe 3, U.S. Not Even Z-Zero."[17]

The first major repercussion of the European triumph was the cancellation of ISABELLE, a proton-proton collider then under construction at Brookhaven. Although hundreds of millions of dollars had already been spent on the project and its tunnel had already been excavated, in July 1983, a subpanel of the High Energy Physics Advisory Panel of the Department of Energy decided that the anticipated energy of the collider, around 400 GeV, would be insufficient to generate new discoveries beyond what had just been found.

With the W and Z identified, the next step would be to find the remaining ingredients of the Standard Model including the top quark, the tau neutrino, and the Higgs. Other goals included finding hypothetical new particles predicted in models seeking to extend the Standard Model into more comprehensive unification schemes.

For example, in the 1970s and 1980s, a number of researchers developed Grand Unified Theories—schemes designed to incorporate QCD along with the electroweak interaction into a single theory. The idea was that at high enough energies, such as in the nascent moments of the Big Bang, all of these interactions would be comparable in strength. With the cooling of the universe, these would bifurcate during two distinct phase transitions into the strong and electroweak interactions and then the strong, weak, and electromagnetic interactions. Thus the original perfect symmetry would break down over time as the vacuum changed its fundamental character.

Even farther reaching unification schemes proposed around that period included supersymmetry, the hypothesized means of uniting fermions and bosons into a comprehensive theory. Each fermion, according to this hypothesis, has a boson counterpart, called its superpartner. Similarly, each boson has a fermion companion. When the universe was a steaming primordial soup, the partners and superpartners were on equal footing, but since then, with the universe cooling, supersymmetry has spontaneously broken—rendering the superpartners too massive to be readily observed. Following a tradition in particle physics of zany nomenclature, the hypothetical boson superpartners of electrons were christened "selectrons" and those of quarks, "squarks." The fermion counterparts of photons were named "photinos," those of gluons, "gluinos," and those of the W and Z, the bizarre-sounding "winos" and "zinos." Theorists hoped signs of the lightest of these superpartners would turn up in collider debris.

As a result of these and other novel theories, by the mid-to-late 1980s, though many of the major predictions of the Standard Model had been verified, no one could complain that there weren't enough projects in high-energy physics to go around. The problem would be cranking out enough juice to spawn the sought-after particles. The SPS collider, at 450 GeV, had quickly reached its limits, with no sign to be found of coveted gems such as the top quark or the Higgs, let alone more exotic particles.

Another CERN project, the Large Electron-Positron Collider (LEP), proved ambitious in size, if not in overall energy. A ring seventeen miles in circumference and hundreds of feet deep, it extended CERN's reach far beyond the Geneva suburbs into the verdant countryside across the French-Swiss border. One of the reasons it was built so large was to reduce the amount of radiation emitted by the electrons and positrons—the greater the radius, the smaller the radiative energy lost.

The LEP's construction required some adjustments to CERN operations. The SPS was adapted to serve as a source for electrons and positrons, which were injected into the LEP ring in countercircling beams before being brought to crash together in a crescendo of energy. By knowing the ring radius, the frequency of the electrons and positrons, and other factors, researchers could calculate the total energy of each collision, allowing for precise determinations of the masses of particles produced.

During its eleven-year run (1989–2000), the LEP was the most powerful *lepton* collider in the world—but, because electrons are so much lighter than protons, lepton colliders are generally weaker than comparably sized hadron colliders. Its energy ranged from just under 100 GeV (when it opened) to slightly over 200 GeV (after upgrades)—insufficient, as it turned out, to find the Higgs or to beat competitors to the top quark. Nevertheless, it was an adept factory for manufacturing W and Z bosons, pinning down their masses with a jeweler's precision.

The top quark would be located among the same Illinois cornfields where the bottom quark was found almost two decades earlier. (The two quarks are sometimes also called "truth" and "beauty.") Nobody expected the wait would be so long and that the second member of the third quark family would be quite so heavy. Its 1995 discovery would be the pinnacle (so far) of the Tevatron's impressive run.

Wilson had stepped down from Fermilab's directorship in 1978, shortly after the upsilon—the first particle known to house a bottom quark—was discovered. After pouring his heart and soul into the Tevatron, he had become incensed when the Department of Energy initially didn't offer enough funding that year to keep it on schedule for a speedy completion.[18] To buttress his argument, he had handed in his resignation and astonishingly it was accepted. Fresh off his bottom quark discovery, Lederman was appointed the new director. He would preside over the Tevatron's opening and chart its course throughout the 1980s. (In 1989, John Peoples became director—succeeded by Michael Witherell and then Pier Oddone. Lederman has maintained a role as director emeritus.)

Physicists around the world celebrated the Tevatron's inauguration, keenly aware that its superior power would offer the best chance to advance scientific understanding of the subatomic realm. On July 3, 1983, twelve hours after its beam was turned on, it reached 512 GeV and broke the world record for energy produced in an accelerator. Among the applauders, Herwig Schopper, CERN's director-general at the time, sent Lederman a gracious message by telex:

> Our warmest congratulations for the extraordinary achievement to accelerate protons for the first time in a superconducting ring to energies never obtained before. Fermilab pioneered the construction of superconducting magnets, opening up a new domain of future accelerators. Please convey our admiration to all the staff concerned.[19]

Though CERN remained Fermilab's major competitor, much of the competition at the Tevatron was internal. With the success of the vying UA1 and UA2 groups as models for a Darwinian approach to discovery, Lederman advocated competing teams at the Tevatron, too. Each would use its own designated detector and analyze its own data. The natural advantage of such a strategy was to subject each group's findings to independent verification by the other.

The first Tevatron team, called the Collider Detector at Fermilab (CDF) Collaboration, consisted of thousands of researchers and technicians from the United States, Canada, Italy, Japan, and China—representing three dozen universities and other institutions. The sheer number of people involved meant that the first few pages of each paper the collaboration published consisted simply of a long list of names.

Still in operation today, the CDF is a multifaceted, hundred-ton device that surrounds one of the beam intersection points and sifts through collision debris for interesting events. As in the case of the SPS, only a minute percentage of all collisions are suitable for analysis. Only if the quark and antiquark constituents of the proton and antiproton beams strike each other directly, interact, and produce debris flying off at large angles, are the collisions worth talking about. Otherwise they are of little note, akin to commuters inadvertently brushing past one another on the way to their trains.

If particle offspring meet the large angle criteria, the CDF, like all complex detectors, subjects them to a barrage of tests. As in traditional devices such as cloud chambers, ionization plays a major role in modern methods for tracking charged particles. Using thin materials technology, a delicate instrument called a Silicon Vertex Tracker sniffs out the subtlest whiff of charge, pinpointing the locations of fleeing subatomic bodies within ten-thousandths of an inch. An immerse superconducting magnet surrounds another unit, called the Central Tracking Chamber, steering charged particles in a way that allows their momenta to be gauged.

Like marines on their first day of boot camp, that's just the beginning of the rigors the particles must face. The next ordeal includes two different energy-trapping devices, electromagnetic and hadronic calorimeters. These force the particles through hurdles (lead sheets and iron plates, respectively), induce jets or showers, and record the energy they sweat off in the process. Those with the endurance to continue through the calorimetry without losing much energy are likely muons and are picked up by the muon tracker just beyond.

Certain initial readings act as triggers, signaling that something significant may be in the works. Immediately, the full data collection machinery kicks in and records everything that can possibly be known about the event—positions, momenta, and energies—tens of thousands of bits of information in many cases. Otherwise, rare, sought-after processes would be buried in an avalanche of mundane decays.

The final step takes place not in the actual detector at all, but rather in the virtual world of computers. Like crime scene detectives, sophisticated software reconstructs what likely happened. Each event of potential interest is dissected—with any missing energy or momentum duly noted. Because energy and momentum are normally conserved, their absence might point to unseen subatomic thieves such as neutrinos. Thus reconstruction offers the only way of filling in the gaps and representing the full picture of what happens during collisions.

Around the Tevatron ring from the CDF another formidable group of experimentalists, called the D0 (pronounced "dee zero") Collaboration, collects vital data with its own detector. Like the CDF it has a tracking system, calorimeters, and muon-detection and triggering systems—with somewhat more emphasis on calorimetry than tracking. Researchers working on D0 projects have hailed from countries around the globe—from Argentina and Brazil to the United Kingdom and the United States.

One of the consistent contributors to the D0 Collaboration is Stony Brook University in New York. As a graduate student there during the early preparatory stages for the project, I witnessed how much testing and calibration takes place to make sure each detector element performs optimally. Calibration involves comparing a piece of equipment's readings to known values. For example, a researcher could calibrate a temperature sensor by seeing if its readings match those of an analogue thermometer. Without calibration, a detector's results could well be flawed—like a scale improperly balanced and indicating the wrong weight. To help calibrate a Cherenkov detector, I recall having to render a room the size of a closet perfectly light-tight, so that only cosmic rays could enter. It took numerous hours in the dark and many layers of duct tape to make sure no stray photons could be seen. That was only one of many thousands of tests performed by many thousands of researchers over many thousands of days before actual runs could even begin. Like raising exotic orchids, high-energy physics certainly requires patience—making the blossoming all the lovelier.

The extraordinary hard work and persistence of the CDF and D0 teams wonderfully paid off when on March 2, 1995, at a specially arranged meeting at Fermilab, both groups reported incontrovertible evidence that they had identified the top quark. Each had released preliminary results earlier—but they wanted to be certain before announcing firm conclusions. Their proof came from evaluating the energy and other properties of leptons and jets produced in numerous candidate events. Based on these values, each team found the mass to be about 175 GeV, the heaviest particle known to date—weighing as much as a gold atom. No wonder it took such a powerful collider to produce it!

A bittersweet aspect of the discovery is that it was announced barely two years after the cancellation of what would have been the next logical step for American physics. What would have been the largest, most powerful collider in the world was axed in a budget-cutting frenzy. High-energy physics in the United States would never be the same.

7

Deep in the Heart of Texas

The Rise and Fall of the Superconducting Super Collider

It was a tragedy, a catastrophe, a scientific Titanic. . . . The Superconducting Super Collider is . . . gone forever.

—HERMAN WOUK, *A HOLE IN TEXAS* (2004)

In the heart of cowboy country, about thirty miles south of Dallas, the town of Waxahachie bears a brutal scar. Where ninety homes once stood among nearby farmland, 135 acres of land lie arid and abandoned. Pocking the soil are seventeen shafts, hundreds of feet deep, now plugged up and useless. A fourteen-mile tunnel, a masterpiece of engineering that now leads absolutely nowhere, forms an arc-shaped gash underground—believed to be slowly filling with water. Less tangible, but even more indelible, wounds have been the injury to the Texas economy and the damage to the American scientific community's morale.

No one imagined, when the Superconducting Super Collider (SSC) project was in the planning stages, that its outcome would be so dreadful. Ironically, it came into being through the demise of another ill-fated project, ISABELLE. At the July 1983 meeting that led to the severing of ISABELLE's lifeline, members of the same federal panel recommended constructing a much higher energy machine. They proposed pursuing a suggestion by Leon Lederman—put forth the previous year at a high-energy workshop in Snowmass, Colorado, for a colossal device with superconducting magnets set in a flat, unpopulated region of the United States. Given CERN's discovery of the weak boson and Fermilab's announcement that it had broken accelerator energy records, it was time to think big.

For particle physicists, building bigger and better machines makes perfect sense. What are land and money compared to the benefits of unlocking nature's deepest secrets? Sure there is always a risk that billions of dollars and years of effort would be wasted, but if we didn't try we would never be able to complete our understanding of the universe. Yet, as certain critics argue, in many other fields of physics (and science in general), important experiments are conducted at far less cost. Some of these have yielded vital societal benefits—for example, the discovery of transistors revolutionized the electronics industry. Is it then worth it to pursue multibillion-dollar projects? The debate over "big science" versus less expensive, "desk-top" experimentation would play a major role in the fate of the SSC.

Envisioned as a 20 TeV collider—much higher energy than either the Super Proton Synchrotron (SPS) or the Tevatron—the SSC was originally known as the "Desertron." According to historians Lillian Hoddeson and Adrienne W. Kolb, there were two main reasons for this designation: "It was thought to require a large expanse available only in the desert, and as it was to penetrate the bleak 'desert' of physical processes described by the theory of grand unification."[1]

The fear was that the newly opened Tevatron would not prove powerful enough to bridge the gap between the final ingredients for the Standard Model and any interesting new physics that lay beyond. No one knew then—and we still don't know—if the Tevatron has the capacity to find the Higgs, let alone even higher energy particles predicted by various Grand Unified Theories and supersymmetric theories. It would be immensely frustrating if elegant new unification theories, linking the electroweak and strong forces in mathematically satisfying ways, could never be physically tested. The frustration would be akin to discerning life on a distant planet but not being able to reach it. Spanning the broad expanse between imagination and investigation would surely require unprecedented technology.

What if the Europeans developed the next generation of colliders themselves, leaving the home of the brave shaking in its boots? Lederman correctly surmised that CERN's seventeen-mile Large Electron-Positron Collider tunnel would be a tempting place to insert a ring of superconducting magnets and create a thunderous proton collider. Best get started on a native machine, American physicists realized, before the Old Country helped itself to a buffet of Nobel Prizes.

In December 1983, the Department of Energy formed a planning group, called the Reference Designs Study (RDS), drawing talented individuals, including directors and staff, from the major accelerator labs in the United States. Representatives from Fermilab, Stanford Linear Accelerator Center, Brookhaven, and Cornell ironed out the preliminary details of how one of the most ambitious scientific projects of all time would be implemented. To be fair to all, the meetings rotated from lab to lab. Cornell physicist Maury Tigner, one of Wilson's former students, headed the group.

By April 1984, the RDS had produced three different options for a proton-proton collider design with 20 TeV of energy per beam and a luminosity of about 10^{34} (1 followed by 34 zeroes) particles per square inch per second. That's about 100,000 times more intense than the SPS beams that spawned the W and Z bosons—like packing the entire population of Topeka, Kansas, into a subway turnstile intended to fit a single person. To accomplish this tight squeeze, each option involved a distinct superconducting magnet design—a strong one with a giant iron yoke and a superconducting coil, a relatively weak one with even more iron (called superferric and producing a smoother magnetic field), and a third, medium-strength one that was iron-free (perhaps they already had too many irons in the fire). In a report to the Department of Energy (DOE), the team concluded that all three designs were feasible and recommended further study.

The next phase in planning the SSC took place through a team called the Central Design Group (CDG), situated at Lawrence Berkeley National Laboratory and also led by Tigner. Several different labs, including the Berkeley Lab, Brookhaven, Fermilab, and the Texas Accelerator Center, a research institute established by collider expert Peter McIntyre, were recruited to investigate the various superconducting magnet designs. The majority of the group supported the strong-field magnet that would require a ring about fifty-two miles in circumference, as opposed to the weak-field design that would necessitate a ring up to a hundred miles long. It would be hard enough to find enough land to support the smaller ring.

By late 1986, the SSC design had been submitted to the DOE and the major labs were united in their support. Because of the projected multibillion-dollar cost for the project, to move forward the approval of President Ronald Reagan, the American leader at the time, was required. On the face of it, that would seem to be a hard sell, given that the federal budget was already stretched through various military and scientific programs. Expensive projects at the time included the Strategic Defense Initiative (more commonly known as "Star Wars") for developing missile-interception systems and a program to establish an international space station. How could the SSC carve out its own piece of a shrinking pie?

Luckily for the prospects of gaining approval, if there was a frontier to be conquered or an enemy to be defeated, Reagan believed in taking major risks. Famously, long before becoming a politician, he played the role of a football player for an underdog team in the film *Knute Rockne, All American*. Like his character George Gipp, better known as "the Gipper," optimism and determination steered his judgments. The SSC was a project on the scientific frontier that represented a means of remaining competitive with Europe. That was the angle its proponents emphasized to win over Reagan's support.

A DOE official asked Lederman if he and his staff could produce a ten-minute video for the president about the questions in high-energy physics the SSC could potentially resolve. To appeal to Reagan's cowboy spirit, Lederman decided to emphasize the frontier aspects of the research. In the video, an actor playing a curious judge visits a lab and asks physicists questions about their work. At the end he remarks that although he finds such research hard to comprehend, he appreciates the spirit behind it, which "reminds [him] of what it must have been like to explore the West."[2]

Apparently the SSC advocates' arguments were persuasive, because Reagan was deeply impressed. Members of his cabinet, concerned about the fallout from the project's asteroidlike impact on the budget, ardently tried to intercept it. With all of their strategic defenses, however, they could not shoot it down.

At Reagan's January 1987 announcement that he would support the SSC, he took out an index card and recited the credo of writer Jack London:

> I would rather be ashes than dust,
> I would rather my spark should burn out in a brilliant blaze,
> Than it should be stifled in dry rot.
> I would rather be a superb meteor,
> With every atom of me in magnificent glow,
> Than a sleepy and permanent planet.[3]

After reading the credo, Reagan mentioned that football player Ken Stabler, known for coming from behind for victory, was once asked to explain what it meant. The quarterback distilled its sentiments into the motto, "Throw deep!"[4] Surely the erstwhile "Gipper"-turned-president would aim no lower. The SSC would move ahead, if he could help it.

At a White House ceremony the following year, Reagan heralded the value of the SSC, saying, "The Superconducting Super Collider is the doorway to that new world of quantum change, of quantum progress for science and for our economy."[5]

Following Reagan's endorsement came the battle for congressional approval, which would not be so easy. Many members of Congress balked at the idea of the United States carrying the ball alone. Consequently, the SSC was marketed as an international enterprise, involving Japan, Taiwan, South Korea, various European nations, and others. *Newsweek* described selling the project this way: "The DOE promised from the start that other nations would help bankroll the SSC. That pledge has greased the project's way through many sticky budgetary hearings."[6]

Significant international support was hard to come by, however. The Europeans in particular were naturally more interested in seeing CERN succeed than in supporting an American enterprise. That's around the time the Large Hadron Collider (LHC) project was first proposed—clearly a higher priority for Europe.

A *New York Times* editorial on May 20, 1988, argued that any American funding toward the SSC would be better spent contributing toward the LHC instead:

> This is a tempting but dangerous initiative because funds to pay for it almost certainly would be stripped from other physics research. . . . The field is of high intellectual interest, and it would be a sad day if the United States did not remain a major player. But European physicists have shown how an existing collider ring at Geneva could be

upgraded to within probable reach of the Higgs boson. Buying into the European ring would be cheaper.[7]

By mid-to-late 1988, Congress had allocated $200 million toward the SSC. Proceeding with caution, it mandated that the money could be used only for planning and site selection. Because an election was imminent, funds for the actual construction would be left to the consideration of the next administration.

Once Congress offered a kind of yellow light for the project to move forward cautiously, the politics behind the venture became even more intense. Many depressed regions salivated over the potential for jobs. Which state would be lucky enough to acquire what the *Times* called "a $6 billion plum"?[8]

To be fair, the competition to find a suitable site was judged by a committee established by the respected and politically neutral National Academy of Sciences and National Academy of Engineering. Of the forty-three proposals submitted from twenty-five states, including seven from the state of Texas alone, the committee narrowed them down to seven. The Lone Star State was clearly the most eager; its government established the Texas National Research Laboratory Commission (TNRLC) well in advance and sweetened the pie by offering $1 billion toward the project from its own budget. What better place to think big than in Texas? Its submission was so hefty—rumored to weigh tons—it was hoisted to the DOE office by truck.[9] In November 1988, after considering the relative merits of the finalists, the DOE announced that Ellis County, Texas—a flat, chalky prairie region with little vegetation except for grasses, shrubs, mesquite, and cactus—was the winning location.

By odd coincidence, the site location announcement occurred just two days after a Texan—then Vice President George Herbert Walker Bush—was elected president. The DOE selection group assured the public that politics played no role in the decision. Through a spokesperson, Bush asserted that he had no involvement in the process and that he found out about the choice when everyone else did.[10]

With the choice of a completely undeveloped site in Texas, many supporters of Fermilab, which had also bid, were left fuming. Fermilab already had much of the infrastructure in place to support an expanded mission; a new location would require starting from scratch with all new staff, buildings, and equipment. In particular, the Tevatron could have been adapted, like the SPS at CERN, to serve as a preaccelerator for the protons entering the collider. "If it was at Fermilab, it would have existed now,"[11] said Brookhaven physicist Venetios Polychronakos, who was involved in planning an SSC experiment.

Some feared a brain drain from Fermilab, with top researchers finding new positions in Texas. Lederman, who had a key stake in both institutions as Fermilab's director and one of the SSC's original proposers, expressed mixed feelings. He anticipated a "certain loss of prestige" for his own lab.[12] However, he expected that it would remain a premier research facility—at least during the years when the SSC was under construction.

Soon after its bid was accepted, the Texas state government, through the TNRLC, assembled a parcel of almost seventeen thousand acres near Waxahachie. For the land of Pecos Bill, that was just a mere pasture. The TNRLC also arranged for environmental studies and administered a research and development program to support the lab. The state's commitment to the project remained steadfast until the end.

The federal commitment to the project was murkier, as there were many clashing forces involved. Congress, the DOE, and research physicists themselves each had different interests, which were sometimes at odds. Although Tigner's CDG performed the bulk of the original planning for the project, when it came time to move forward with constructing the collider and setting up the lab, the group was bypassed in favor of the Universities Research Association (URA)—a consortium that managed Fermilab and with which the DOE felt comfortable working. Tigner was seen as a "cowboy in the Wilson tradition"[13] and as potentially having difficulty bending to the demands of Congress and the DOE. Thus, the URA instead chose a relative newcomer, Harvard physicist Roy Schwitters, to become lab director. After serving briefly under Schwitters as deputy director, Tigner stepped down in February 1989 and returned to Cornell. Sadly, with his resignation, the construction of the SSC would need to proceed without his vital technical experience, including the five years he spent helping design the accelerator.

As scientific historian Michael Riordan has pointed out, Schwitters and the URA sharply departed from prior practice by integrating private industrial contractors into the decision-making process.[14] Before the SSC, accelerator laboratories were planned solely by research physicists—who would employ technicians if needed for particular tasks. As we've seen, for instance, Wilson designed almost all aspects of Fermilab himself. Schwitters had a different philosophy, involving industrial representatives as well as academics to solve the engineering challenges associated with building what would have been the world's most formidable scientific apparatus. Among these were corporations heavily involved in the defense and aerospace industries, and for whom it was their first exposure to high-energy physics. Many of the industrial workers switched jobs because of military cutbacks due to the end of the cold war. Because they were used to a certain mind-set, the lab assumed some of the secretive aspects of a defense institute. Moreover, some of the established physicists felt that they couldn't handle the scientific demands. These factors, as Riordan noted, created a clash of cultures that alienated many of the experienced researchers and made it hard to recruit new ones.

Despite these internal tensions, during the first Bush administration, the SSC benefited from strong federal support. Familiar with the Texas landscape due to his years as a businessman and politician in that state, President Bush considered the SSC a national scientific priority and pressed Congress each year to fund it. The DOE distributed particular tasks such as building and testing detectors throughout almost every state, offering politicians around the county further incentive to favor it.

The program even survived a doubling of its estimated completion cost to a whopping $8 billion, announced in early 1990. This came about when engineering studies forced a major redesign of the project due to issues with the magnets and other concerns.

Superconducting magnets are in general extremely delicate instruments. The stronger a magnet's field, the greater its internal forces and the higher the chance that its coil and other parts will subtly oscillate. Vibrations cause heating, which can ruin the superconducting state and weaken or destroy the magnets. At 6.5 Tesla (the metric unit of magnetism)—more than 50 percent stronger than the Tevatron's field—the SSC's magnets were very much at risk. To minimize tiny movements, the magnets included carefully placed steel clamps. Getting them to work effectively was a matter of trial and error. In an important preliminary test of twelve magnets, only three passed muster. Designers struggled to improve the performance.

Another question concerned the size of the magnets' openings. Smaller apertures were cheaper but entailed a greater risk to the streams of protons passing through. Any misalignment could reduce the rate of collisions and sabotage the experiment. Ultimately, after considerable discussion, the SSC administration decided to enlarge the magnet openings to allow more room for error.

Other design changes made at that time included increasing the ring circumference to fifty-four miles and doubling the proton injection energy (the energy protons are accelerated to before they enter the main ring). All of these modifications forced the bill sky high. Although some members of Congress became livid when learning about the huge increase in the cost, the general reaction at that time was to increase oversight rather than close down the project. Construction funds started to flow, and the lab began to take shape starting in the fall of 1990.

As planned, the SSC was to have a succession of accelerators boosting protons to higher and higher energies before they would enter the enormous collider ring. These consisted of a linear accelerator and three synchrotrons of increasing size: the Low Energy Booster, the Medium Energy Booster, and the High Energy Booster. Tunnels for the linear accelerator and the smallest synchrotron were the first to be excavated.

Boxy structures, to support future operations underground, sprang up like prairie grass along the flat terrain, including the Magnet Development Lab, the Magnet Test Lab, and the Accelerator Systems String Test Center. These were facilities for designing, building, and testing the various types of superconducting magnets needed for the project. Two large companies, General Dynamics and Westinghouse, took on the task of building the thousands of dipole magnets that would steer the protons, like roped-in cattle, around what would have been the largest underground rodeo in the world.

Meanwhile, drawn by the prestige of contributing to a kind of Manhattan Project for particle physics, or simply finding a good-paying position, more than two thousand workers relocated to Texas. The lure of potentially finding the Higgs boson or supersymmetric companion particles enticed many an adventurous physicist to venture south of glittery Dallas and try his or her luck with collider roulette. For researchers who already had thriving careers, it was a significant gamble. Some took leave from their full-time positions; others gave up their old jobs completely with hopes of starting anew.

In the tradition of colliders supporting a pair of major detectors, with collaborations lined up behind these, two groups' proposals were approved for the SSC. The first, called the SDC (Solenoidal Detector Collaboration), involved almost a thousand researchers from more than a hundred institutions worldwide and was headed by George Trilling of the Berkeley Lab. Its general-purpose detector was designed to stand eight stories high, weigh twenty-six thousand tons, and cost $500 million. The target date for it to start collecting data was fall 1999—offering hope of finding the Higgs before the toll of the millennial bells.

The second group, called GEM (Gammas, Electrons, and Muons), was led by Barry Barish of Caltech, an accomplished experimentalist with a stately beard and shoulder-length silver hair, along with liquid argon calorimetry coinventor William Willis of Columbia University, and a humongous cadre of researchers. Their project involved a detector specially fashioned for pinpoint measurements of electrons, photons, and muons. GEM was supposed to be located around the ring from the SDC at a different intersection point, collecting data independently, like competing newspapers housed in separate city offices.

Unfortunately, neither of these detectors ever had a chance to taste flavorful particles. As the SSC project rolled through the early 1990s, it accumulated more and more opposition—not just from politicians dismayed that it would break the budget but also from fellow physicists in fields other than high energy. Most branches of experimental physics don't require $8 billion budgets, yet can still yield groundbreaking results.

Take for example high-temperature superconductivity. In the 1980s, Swiss physicist Karl Müller and German physicist Johannes Bednorz, working with reasonably priced materials in the modestly sized (compared to CERN or Fermilab at least) complex of IBM's Zurich Research Laboratory, revolutionized physics with their discovery of a ceramic compound that could conduct electricity perfectly at temperatures higher than previously known superconductors. Other experiments in various labs, including work by Paul Chu at the University of Houston, turned up substances with even higher transition temperatures. Although these ceramic superconductors still need to be quite cold, some maintain their properties above the temperature of liquid nitrogen. Immersing a material in liquid nitrogen is far cheaper than the drastic methods used to create the near-absolute-zero superconductors once believed to be the only types. Therefore not only did Müller and Bednorz's finding come with a much cheaper price tag than, say, the top quark, it also led to cost saving for future research and the potential for more widespread applications of superconductivity.

Because discoveries related to material properties bear more directly on people's lives than does high-energy physics, many researchers in these fields, such as Cornell physicist Neil Ashcroft, have argued that they deserve at least as much support. "Things are out of whack," he said. "Condensed-matter physics is at the heart of modern technology, of computer chips, of all the electronic gadgets behind the new industrial order. Yet relative to the big projects, it's neglected."[15]

Another leading critic of "big science," who was skeptical about channeling so much funding into the Super Collider, was Arno Penzias, codiscoverer of the cosmic microwave background. Penzias said, "One of the big arguments for the S.S.C. is that it will inspire public interest in science and attract young people to the field. But if we can't educate them properly because we've spent our money on big machines instead of universities, where's the point? As a nation we must take a new look at our scientific priorities and ask ourselves what we really want."[16]

On the other hand, who could anticipate what would have been the long-term spin-offs of the SSC? In the past, some discoveries that seemed very theoretical at the time, such as nuclear magnetic resonance, have ended up saving countless lives through enhanced techniques for medical imaging and treatment. But since nobody had a crystal ball for the SSC and its potential applications, its critics painted it as just big and expensive.

The rising crescendo of arguments against "big science" and in support of smaller, less expensive projects jived well with growing congressional sentiments that the SSC was getting out of hand. Given that Congress was promised that substantial foreign contributions would fill out the SSC's budget, when these failed to materialize, many members were understandably miffed. Some didn't think that Schwitters and the DOE under Secretary of Energy James Watkins were managing the project effectively. Still, it came as a surprise when in June 1992 the House of Representatives voted 232–181 in favor of a budget amendment that would end the project.[17] Only the Senate's support for the SSC temporarily kept it alive.

In the spirit of former senator William Proxmire's "Golden Fleece Awards" for alleged government boondoggles, many of those who favored terminating the SSC painted it as a wasteful endeavor that would benefit only a small group of eggheads. In times of tight budgets, they wondered, why channel billions of dollars into crashing particles together to validate theories rather than, say, blasting away at the all-too-real federal deficit behemoth?

"Voting against the SSC became at some point a symbol of fiscal responsibility," said its then associate director Raphael Kasper, who is currently vice president of research at Columbia. "Here was an expensive project that you could vote against."[18]

In January 1993, Bill Clinton succeeded Bush as president. Without the Texas connection, a key strand of the SSC's support dissolved. Although Clinton indicated that he backed the project, particularly in a June letter to the House Appropriations Committee, he advocated extending the time line for three extra years to reduce the annual impact on the federal budget. Postponing the SSC's targeted opening date (to 2003) made it seem an even riskier venture, however, because it could well have been obsolete by the time it went on line. What if the Tevatron had found the Higgs boson by then?

Once the collider lab's anticipated costs rose to approximately $10 billion, largely because of the pushing back of its schedule, it was only a matter of time before an increasingly frugal Congress signed a do-not-resuscitate order. A House of Representatives vote on October 19, 1993, denied by a two-to-one margin a funding bill that would have supported further construction. Instead, the SSC's annual appropriation was directed to mothball the part of the facility that had already been built. By then $2 billion had already been spent and more than one-quarter of the project was complete—all for naught. The tangible result of a decade of planning and hard work would just be boarded up and shrouded in dirt. Requiescat in pace.

The cancellation of the SSC did, in the short term, save federal money. Along with many other cost-cutting measures, the federal budget would be balanced by the end of the decade. (Ironically, in the 2000s, the deficit would skyrocket again, making all of the cost cutting moot!) Yet, what is the long-term price of a national decline in scientific prestige? Skipping the moon landings, eschewing the robotic exploration of Mars, and abstaining from telescopic glimpses at the swirling mists of ancient galaxies would have each cut government expenses, too—while extinguishing the flames of our collective imagination. If it is a choice between science and sustenance, that's one thing, but surely our society is rich enough to support both. It remains to be seen whether the United States will ever resume its pioneering mantle in high-energy physics. Thus in retrospect, many see the abandonment of what would have been the premier collider in the world as a grave error.

According to Fermilab physicist William John Womersley, "The SSC has cast a very long shadow over high-energy physics and big science in general. We're still dealing with the legacy."[19]

In the aftermath of the closure, those who took the career risk and moved down to Texas for the SSC met with varying degrees of disappointment. Some regrouped, sent out their résumés (or were recruited), and managed to find new positions in other labs or universities. For the experienced physicists, finding an academic position was hard, because not many universities wished to hire at the senior level, and the closure of the SSC reduced the need for professors in the high-energy field. A survey taken one year after the closure found that while 72 percent of those in the SSC's Physics Research Division had found employment, only 55 percent of those positions were in high-energy physics.[20]

Other workers, who had laid down deep roots in Texas and didn't want to leave, either found other types of jobs or simply retired early. A few stayed to help sell off the equipment and assist in attempts to convert the site to alternative uses.

Given all of the time and energy that went into assembling the land, digging the tunnels, and constructing the buildings, it is remarkable that the site has yet to be put to good use. The federal government transferred the property to the state of Texas, which in turn deeded it to Ellis County. For more than fifteen years, the county has tried in vain to market the structures, particularly the former Magnetic Development Laboratory. Like Dickens's forlorn spinster, Miss Havisham, the relic building is a jilted bride frozen in time with no interested suitors. An agreement to convert it into a distribution center for pharmaceutical products fell through, and informal plans to house an antiterrorism training base never materialized. In 2006, trucking magnate J. B. Hunt's plans to use it as a data center were abandoned upon his death.[21] It did, however, play a background role in a straight-to-video action flick, *Universal Soldier II*.[22] To mention another has-been, the Norma Desmond of labs finally had its close-up.

Though it's instructive to ponder what could have been in hypothetical scenarios about alternative choices, in truth physicists can't afford to wallow in disappointment. An energetic frontier is ripe for exploration and there's no time for looking back. Leaving the plains and pains of Texas behind, in the late 1990s the American particle physics community regrouped and headed either north to Illinois, for renewed efforts at the Tevatron, or across the ocean to the cantoned land where cubed meat and melted cheese deliciously collide. After all, Geneva, Switzerland, has distinct charms, some of which Lederman described well. Comparing it to Waxahachie, he wrote, "Geneva . . . has fewer good rib restaurants but more fondue and is easier to spell and pronounce."[23]

Humanity's best chance of finding the Higgs boson and possibly identifying some of the lightest supersymmetric companion particles now rests with the Large Hadron Collider. Though it will crash particles together at lower energies than the SSC was supposed to—14 TeV in total instead of 20 TeV—most theoretical estimates indicate that if the Higgs is out there the LHC will find it. If all goes well, modern physics will soon have cause for celebration.

8

Crashing by Design

Building the Large Hadron Collider

The age in which we live is the age in which we are discovering the fundamental laws of Nature, and that day will never come again

—RICHARD FEYNMAN *(THE CHARACTER OF PHYSICAL LAW*, 1965*)*

Compared to the wild flume ride of American high-energy physics, CERN has paddled steadily ahead like a steamboat down the Rhone River. Each milestone has been part of a natural progression to machines of increasing might—able to push particle energies higher and higher. While American high-energy physics has become increasingly political—rising or falling in status during various administrations—the independence of CERN's directorship and its commitment, cooperation, and collaboration to carrying out projects already proposed have enabled it to successfully plot the laboratory's course for decades ahead.

One aspect of CERN's impressive efficiency is its ability to recycle older projects into key components of state-of-the-art devices. The old Proton Synchrotron, upon its retirement as a stand-alone machine, became an injector for the Super Proton Synchrotron (SPS). The SPS, in turn, has been used for a variety of purposes, including serving as a preaccelerator for more powerful devices. Little at CERN ever truly goes to waste, and this keeps costs relatively low.

This tendency to adapt obsolete projects for reuse as parts of new ones reflects the European need to conserve space and vital resources. Europe is more crowded and doesn't have the luxury of unbridled development. Therefore a venture like the SSC involving building a completely new facility from scratch in a region far from other labs would be much less likely to happen.

By making use of the old seventeen-mile tunnel for the Large Electron-Positron Collider (LEP), the Large Hadron Collider (LHC) serves as the perfect example of accelerator recycling. Digging the LEP tunnel was a colossal undertaking. From 1983 to 1988, it represented the largest civil-engineering project in Europe. Because the main ring had to be wedged between the Geneva airport and the Jura mountains, engineers had little room to maneuver. Tunnel diggers were forced to blast through thick layers of solid rock. To account for changes in topography, the ring had to be tilted by one and a half degrees. Amazingly, the tunnel lined up nearly perfectly (when its ends were joined, they were less than half an inch off) and was precisely the right size. It is therefore fortunate that the LHC hasn't required a whole new tunnel but rather could be fit into the old one.

The decision to drop ultracool superconducting magnets into the LEP ring and turn it into a hadron collider—at the suggestion of Rubbia and others—was first discussed in the 1980s. (Hadrons, such as protons, are much more massive than electrons and thereby require far stronger magnets, such as superconducting magnets, to be steered through the same ring.) Reportedly, practically from the time the LEP opened, Rubbia was anxious to have the tunnel converted. While the SSC was under construction, many CERN physicists hoped that the LHC could be finished first. A frequent rallying cry was to plan on opening the LHC two years before the SSC—thus beating the Americans to the good stuff. The SSC's cancellation added further impetus to the project, as it meant that the LHC would represent the main—or only—hope for finding certain massive particles. The model of international competition and vying labs confirming one another's results, which served well during an earlier era of smaller machines, would need to be replaced with international cooperation, centered in Europe.

The final decision to build the LHC came little more than a year after the SSC was canned. On December 16, 1994, CERN's nineteen member nations at the time voted to budget $15 billion over a two-decade span to build what would be the world's mightiest collider. Through its leaders' firm commitments, the continent that spawned Galileo and Kepler readied itself to be the vanguard of science once more.

Unlike with the SSC, politics played little discernable role in the LHC's construction. Each European country that belongs to CERN contributes a specific annual amount that depends on its gross national product. Richer countries, such as Germany, France, and the United Kingdom, shell out the bulk of CERN's budget—generally with no yearly debate over how much that amount will be. (The United Kingdom has recently become more cautious about future scientific projects, however.) Thus CERN administrators can rely on certain figures and plan accordingly.

Furthermore, unlike the American case, in Europe regional competitions have never decided where projects are built. The French communities of the Pays de Gex, the region where much of the tunnel is located, didn't launch a "Don't Mess with Gex" campaign to sway politicians one way or the other. Rather, they've quietly accepted CERN as a long-standing neighbor that shares the land with farmers, wine growers, cheese makers, and other producers. As the region's motto proclaims, Gex is *"Un jardin ouvert sur le monde"* (a garden open to the world).

On the Swiss side, Geneva is used to all manner of international enterprises. The city where the League of Nations was established and famous treaties were signed now houses a plethora of global organizations—the U.N. European headquarters, the World Health Organization, the International Labour Organization, the International Federation of Red Cross and Red Crescent Societies, and many others. CERN is well accepted along with its kindred cooperative institutions. The medley of researchers' foreign languages—including English, Russian, and field theory—is nothing special; Genevese diplomats can match that Babel and more.

Furthermore, over the centuries Geneva has seen its share of groundbreaking movements. Compared to the impact of the Reformation and the Enlightenment, slamming particles together underground barely registers on the city's Richter scale of history.

True, the French countryside west of Geneva is much quieter. To ensure a harmonious relationship, CERN has sought to minimize its impact on that region. The patchwork of pastures and vineyards that front the misty Jura mountains displays no visible indication that a giant particle-smashing ring lies hundreds of feet beneath them. Only the occasional road signs steering CERN vans to scattered laboratory buildings, and the power lines scratching through the green and golden tapestry, offer clues as to what lies below.

The latter represents possibly the biggest source of contention—CERN is a huge drain on the region's electricity. Originally, this electricity was supplied by Switzerland; now it is furnished by France. When its machines are fully running, CERN expends about as much power as the entire canton of Geneva. Because of the predominance of electrical heating in the area, this usage would be felt mainly during wintertime. Consequently, as a considerate neighbor, CERN often adjusts its power needs to accommodate—for example, by scheduling shutdowns during the coldest time of year. Though this means less data collection, fortunately for the more athletic researchers, the winter closures are timed well with peak skiing season in the nearby Alps.

To prepare for the LHC project, the LEP tunnel needed to be completely gutted. After the final LEP runs took place in 2000, the refurbishing could finally begin. Orders went out for thousands of superconducting magnets of several different types. One kind, called dipole magnets, were designed to steer twin proton (or ion) beams around the loop. (A subset of the LHC experiments will involve accelerating ions rather than protons.) Dipoles tend to guide charged objects in a direction perpendicular to their magnetic fields—ideal for maneuvering. A second variety of magnets, called quadrupoles, were targeted at focusing the beams, to prevent them from spreading too much. To simplify the LHC's design, these were placed at regular intervals. Other more complex magnet designs—called sextupoles, octupoles, and decupoles—were added to the mix to provide finer beam corrections. Like a delicate space mission, the orbit needed to be tuned just right.

Because the particles rounding the LHC would be alternatively steered and focused, with ample room for experimentation, the machine was not planned to be a perfect circle. Rather, it was divided into eight sectors, each powered separately. Sectors consist of curved parts and straight intervals—the latter used for a variety of purposes including injecting particles, narrowing the beams, and conducting experiments.

Researchers realized that two extreme conditions would need to be maintained to make the LHC a success. These requirements would bring some of the most hostile aspects of outer space down to Earth. First, the twin beam pipes, riding through the apertures of the magnets, would need to be kept as close to vacuum states as possible. That would allow the protons (and ions) to reach ultrahigh energies without bouncing off of gas molecules as in a pinball game. A pumping system was chosen that would maintain the pressure at 10^{-13} (one-tenth of one trillionth) that of the atmosphere at ground level. That's far from as empty as the interplanetary void, but it's closer to a pure vacuum than virtually anywhere else on Earth.

Second, the thousands of magnets would need to be supercooled well below the critical temperatures that maintain their superconducting states. This would keep their magnetic fields as high as possible—more than 8.3 Tesla, double the field used at the Tevatron. To keep the temperatures so low would require superfluid helium—a highly correlated ultracool state of that element—at 1.9 degrees Kelvin (above absolute zero). That's even colder than the microwave background radiation detected by Penzias and Wilson in their confirmation of the Big Bang.

At first glance, it would seem to be prohibitively expensive to keep so many magnets so cold. Indeed, superfluid helium is very costly to produce. However, by surrounding each "cryomagnet" (as supercooled magnets are called) with an insulating vacuum layer, little heat from the outside would leak through. Emptiness is a great thermal blanket.

Another factor LHC designers had to reckon with has to do with lunar influences. Remarkably, the moon has a periodic lure on the region. No, there aren't full-moon-crazed werewolves haunting the woods near Ferney-Voltaire and Meyrin, eager to rummage through supercooled containers looking for frozen steaks—at least as far as we know. Rather, the moon's effect is purely gravitational. Just as it pulls on the oceans and creates the tides, the moon tugs on the ground, too. Rocks are certainly not as pliable as water, but they do have a degree of elasticity. Due to its lunar stretching, Earth's crust in Geneva's vicinity rises and falls almost ten inches each month. This creates a monthly fluctuation in the LHC's length of about 1/25 of an inch.[1] The effect was first noted when the tunnel was used for the LEP and has been accommodated through corrective factors in any calculations involving ring circumference.

Topography played an even more important role when it came time to equip the LHC with detectors. Completely new caverns were excavated, with the largest, at "Point 1," to accommodate the most sizable detector, ATLAS (A Toroidal LHC ApparatuS). Three other detectors, called CMS (Compact Muon Solenoid), ALICE (A Large Ion Collider Experiment), and LHCb (Large Hadron Collider beauty) were placed at additional points around the ring. The designs for each of these detectors took many years of planning. Their approval recognized their complementary roles in the overall LHC mission—each contributing a unique means of measuring particular types of collision by-products and thereby primed for different kinds of discoveries.

The ATLAS project has been in the planning for more than a decade. It represents a fusion of several earlier projects involving researchers from a number of different countries. Experiences at earlier collider projects—those completed along with those aborted—played a strong role in shaping the detector's design.

Take, for instance, ATLAS's electromagnetic calorimetry (energy-gauging) system. It relies on a method William Willis proposed in 1972 for the ill-fated ISABELLE collider: using liquid argon to convert radiation into measurable electrical signals through the process of ionization. When ISABELLE was canceled, Willis included liquid argon calorimetry again in the proposal he developed with Barish and others for the GEM detector at the SSC. In addition to Brookhaven, where Willis was based, the technique came to be used at laboratories such as Fermilab and SLAC. Now Willis is the U.S. project manager for ATLAS, where his liquid argon method forms a key component of the detector's energy-measuring system.

If liquid argon is the blood flowing through the heart of ATLAS, silicon pixels and strips (wafers responsive to light, like digital cameras) offer the ultrasensitive eyes. Immediately surrounding its interaction point is a zone of maximum surveillance called the inner detector—where electronic eyes gaze virtually everywhere like a particle version of Big Brother. Except for the places where the beam line enters and leaves, the inner detector is completely surrounded by tiny light probes. In other words, it is hermetic, the ideal situation for high-energy physics where virtually all bases are covered. This state of maximum spy-camera coverage offers the optimal chance of reconstructing what happens in collisions.

To encapsulate the beam line in a symmetric way, most sections of ATLAS, the inner detector included, are arranged in a set of concentric cylinders, called the barrel, framed at the entrance and exit by disks perpendicular to the beam, called the end-caps. This geometry means that almost every solid angle from the beam line is recorded. The inner detector's tracking system includes photosensitive pixels and strips covering the three interior layers of the barrel as well as the end-caps.

Between the inner detector and the calorimeters is a solenoid (coil-shaped) superconducting magnet with a field of approximately 2 Tesla. Cryostats (systems for supercooling) keep the magnet at less than five degrees above absolute zero. The purpose of the solenoid magnet is to steer charged particles within the inner detector—bending them at angles that depend upon their momenta (mass times velocity). Therefore, the electronic tracking system, in tandem with the magnet, enables researchers to gauge the momenta of collision products.

After particles breach the boundary of the inner detector, they enter the realm of the electromagnetic calorimeter. Bashing into lead layers, the electromagnetically interactive particles decay into showers and deposit their heat in the liquid argon bath, producing detectable signals. Delicate electronics pick up the signals from all of the energy lost, offering another major component of event reconstruction. Discerning the charge, momentum, and energy of a particle is like asking a soldier his name, rank, and serial number. Because each of these physical quantities is conserved, identifying each particle's information optimizes the chances of figuring out which unseen carriers (such as neutral particles) might be missing.

Only some of the featherweight particles, such as electrons, positrons, and photons, are knocked out completely in the electromagnetic calorimeter; heavier (and nonelectromagnetic) particles can slip through. These bash into a thick layer of steel tiles interspersed with scintillators—the hadron calorimeter. Sensors abutting that layer record the heat deposited by any particles subject to the strong force. There protons, neutrons, pions, and their hadronic cohorts make their final stands.

The only charged particles that can evade both types of calorimeters without being absorbed are muons. To ensnare them, the outermost, and largest, layers compose the muon system. It operates in some ways like the inner detector, with magnets and a tracking system, only on a far grander scale—dwarfing the rest of ATLAS. Pictures of ATLAS taken after its completion inevitably showcase the muon system's colossal end-cap: the Big Wheel.

The muon system's enormous superconducting magnets have a much different shape from the central magnet. Rather than a solenoid, they are toroidal (doughnut-shaped) but stretched out. At one-quarter the length of a football field, they are the largest superconducting magnets in the world. Eight of them carve through the outer barrel—like an eight-way apple slicer. The sheer size of these magnets magnifies the bending of muons as they pass through. Thousands of sensors track the muons' paths as they swerve—revealing those particles' precise momenta.

The particles that survive the full range of detection systems are those that are insensitive to both the electromagnetic

A view of the ATLAS detector with its eight prominent toroidal magnets.

and strong interactions. The prime suspects among these are neutrinos. Because they interact solely through the weak and gravitational forces, neutrinos are very hard to detect. ATLAS does not make an attempt to catch these; rather, components of their momenta and energy are estimated through a subtraction process. Because the protons, before colliding, are traveling along the beam line, their total transverse (at right angles to the beam direction) momentum must be zero. According to conservation principles, the total transverse momentum after the collision—determined by adding up the momenta of everything detected—ought to be zero as well. If it isn't, then subtracting that sum from zero yields the transverse momenta of unseen collision products. Therefore, the ATLAS researchers have a good idea of what the neutrinos have carried off.

Halfway around the LHC ring, beneath the village of Cessy, France, is the other general-purpose detector, CMS. The "compact" in its name reflects the CMS's aspiration to pursue similar physics to ATLAS with a detector a fraction of the volume—although still bigger than a house. Instead of an assortment of magnets, CMS is constructed around a single colossal superconducting solenoid (coil-shaped magnet) that puts out a field of 4 Tesla—approximately a hundred thousand times greater than the Earth's. It surrounds the detector's central silicon-pixel tracker and calorimeters, bending the routes of charged particles within those regions and extracting precise values of their momenta. Knowing the momenta helps the researchers reconstruct the events and deduce what might be missing, such as neutrinos.

Another difference between CMS and ATLAS concerns the way they force electromagnetically sensitive particles to "take a shower." Instead of frigid liquid argon, the CMS electromagnetic calorimeter includes almost eighty thousand lead tungstate crystals (energy-sensitive materials) to measure the energies of electrons, positrons, and photons in particle showers. The hadrons encounter dense curtains of brass and steel, while muons are caught up in layers of drift chambers and iron that lie just beyond the magnet.

The CMS detector before closure.

The two collaborations have much in common: large teams of researchers from institutions around the world, ambitious goals, and the powerful data-capturing technologies required to carry out these bold objectives. Data from the millions of events recorded by each group—those passing the muster of the trigger systems designed to weed out clearly insignificant occurrences—will be sent electronically to thousands of computers in hundreds of centers around the world for analysis by means of a state-of-the-art system called the Grid.

Each team has an excellent shot at identifying the Higgs boson, assuming its energy falls within the LHC's reach. If one team finds it, the other's efforts would serve as vital confirmation. The research paper making the important announcement would literally contain thousands of names. Because of the shared credit, the Nobel committee would be hard-pressed to award its prize to an individual or small set of experimentalists. Unlike, for instance, Rubbia and Van der Meer's winning science's highest honor for the weak boson discoveries, there probably wouldn't be obvious hands (aside from those of its namesake theorist) to confer the award.

Completing the quartet at the LHC's interaction points are two sizable specialized detectors: the LHCb (Large Hadron Collider beauty) experiment and ALICE (A Large Ion Collider Experiment). Two other petite detectors will operate near the ATLAS and CMS caverns, respectively: the LHCf (Large Hadron Collider forward) and TOTEM (TOTal Elastic and diffractive cross-section Measurement) experiments.

The focus of the LHCb experiment is to produce B-particles (particles containing the bottom quark) and to examine their

LHC Detectors and Their Missions

Detector	Full Name	Mission
ATLAS	A Toroidal LHC ApparatuS	General Purpose (including searching for the Higgs boson and supersymmetry)
CMS	Compact Muon Solenoid	General Purpose (including searching for the Higgs boson and supersymmetry)
ALICE	A Large Ion Collider Experiment	Creating quark-gluon plasma
LHCb	Large Hadron Collider beauty	Searching for CP violation in B-particle decay
LHCf	Large Hadron Collider forward	Testing cosmic ray detection devices
TOTEM	TOTal Elastic and diffractive cross-section Measurement	High-precision measurements of protons

modes of decay. B-particles are extremely massive and would likely have a rich variety of decay products that could possibly furnish evidence of new phenomena beyond the Standard Model. In particular, the LHCb researchers will be looking for evidence of what is called CP (Charge-Parity) violation. CP violation is a subtle discrepancy in certain weak interactions when two reversals are performed in tandem: switching the charge (from plus to minus or minus to plus) and flipping the parity (taking the mirror image). Switching the charge of a particle makes it an antiparticle, which does not always behave the same in weak decays. Reversing the parity, as Lee and Yang demonstrated, similarly does not always yield the same results in weak decays. Physicists once believed that the combination of the two operations would always be conserved. However, in 1964, American physicists James Cronin and Val Fitch demonstrated that certain kaon processes subtly break this symmetry. Particular B-meson decays involving the weak interaction also violate CP symmetry—processes that the LHCb experiment hopes to study.

Unlike ATLAS and CMS, the LHCb detector does not surround its whole interaction point. Instead, it consists of a row of subdetectors in the forward direction. The reason is that the B-particle decays to be studied generally fan out in a cone in front of the collision site. Several hundred researchers from more than a dozen countries are members of the LHCb collaboration.

ALICE is an experiment that involves the collision of lead ions rather than protons. The LHC will circulate ions for one month each year to accommodate this project. When the lead ions collide, the hope is that they will produce a state of matter called quark-gluon plasma—a free-flowing mixture of hadron constituents thought to resemble the primordial broth that filled the very early universe. Normally, quarks are confined to hadrons—grouped in pairs or triplets and strung together by gluons. However, under the energetic conditions of the LHC, equivalent to more than a hundred thousand times the temperature at the Sun's core, physicists think such barriers would crumble—liberating the quarks and gluons. This freedom would be extraordinarily brief. The massive detector used to record the outcome has a layered barrel design with eighteen components, including various types of tracking systems and calorimetry. More than a thousand physicists from more than a hundred different institutions are contributing to the project.

The LHCf experiment, the smallest at the LHC, makes good use of some of the leftovers from ATLAS. Standing in the beam tunnel about 460 feet in front of the ATLAS collision point, it is intended to measure the properties of forward-moving particles produced when protons crash together. The goal is to test the capability of cosmic ray measuring devices. Several dozen researchers from six different countries are involved in the experiment.

Finally, TOTEM, a long, thin detector connected to the LHC beam pipe, is geared toward ultrahigh-precision measurements of the cross-sections (effective sizes) of protons. Located about 650 feet away from the CMS detector, it consists of silicon strips situated in eight special vacuum chambers called Roman pots. These are designed to track the scattering profiles of protons close to the beam line. TOTEM involves the work of more than eighty researchers, associated with eleven institutions in eight different nations.

To monitor the progress of the LHC experiments, members of each group conduct regular meetings. Particularly for the larger detectors, each instrumental component requires calibration and careful monitoring. Group members frequently apprise one another of the results of this testing to troubleshoot any potential problems.

One issue that sometimes arises with the more complex detectors is anticipating how one component might affect another's results—for example, through electronic noise. The presence of ultrastrong magnetic fields further complicates matters, as they could exert disruptive influences. During testing at ATLAS in November 2007, for example, one of the toroid magnets wasn't properly secured and it moved about an inch toward an end-cap calorimeter. Fortunately, there was no damage. If a problem is found within one of the hermetically sealed sections, often nothing can be done until it is unsealed and opened up. Typically, such opportunities arise when the LHC is temporarily shut down—close to the winter holidays, for example.

The "machine people," those involved with the planning and operations of the accelerators, have their own separate meetings. Their primary concern is that the overall system is working smoothly. One of the trickiest issues they face is keeping the dipoles, quadrupoles, and other ring magnets at their optimal fields with maximum energies.

If the magnetic fields and energies are raised too high in too rapid a manner, an adverse phenomenon called quenching occurs. Quenching is when part of a superconducting magnet overheats because of moving interior components and destroys the superconductivity. At that point, the magnet becomes normally conducting and its field drops to unacceptable levels. To combat such a ruinous situation, the magnetic fields are ramped up slowly, then reduced, again and again, in a process called training. It's a bit like placing your feet in a hot tub, pulling them out, then putting them back in again, until you are used to the heat.

The detector and the machine groups are well aware of the LHC's limitations. Every machine has structural limits—for example, upper bounds on the beam luminosity due to the magnets' maximum focusing power. Consequently, researchers take note and plan for upgrades well in advance. It is remarkable that while some team members of the various collaborations are readying current experiments, others are involved in developing scenarios for modifying aspects of the detectors and accelerators years in advance. A planned luminosity upgrade to turn the LHC into the "Super LHC" is already intensely under discussion. Modern particle physics requires envisioning situations today, tomorrow, and decades ahead—sometimes all wrapped together in the same group meetings.

Amid all of these preparations, researchers try to keep their eyes on the big picture. Results could take years, but the history of science spans the course of millennia. The identification of the Higgs boson and/or the discovery of supersymmetric companion particles could shape the direction of theoretical physics for many decades to come. Another field eagerly awaiting the LHC findings is astronomy. Astronomers hope that new results in particle physics will help them unravel the field's greatest mystery: the composition of dark matter and dark energy, two types of substances that affect luminous material but display no hint of their origin and nature.

9

Denizens of the Dark

Resolving the Mysteries of Dark Matter and Dark Energy

I know I speak for a generation of people who have been looking for dark-matter particles since they were grad students. I doubt . . . many of us will remain in the field if the L.H.C. brings home bad news.

—JUAN COLLAR, KAVLI INSTITUTE FOR COSMOLOGICAL PHYSICS (*NEW YORK TIMES*, MARCH 11, 2007)

There's an urgency for LHC results that transcends the ruminations of theorists. For the past few decades, astronomy has had a major problem. In tallying the mass and energy of all things in the cosmos, virtually everything that gravitates is invisible. Luminous matter, according to current estimates, comprises only 4 percent of the universe's contents. That small fraction includes everything made out of atoms, from gaseous hydrogen to the iron cores of planets like Earth. Approximately 23 percent is composed of dark matter: substances that give off no discernable light and greet us only through gravity. Finally, an estimated 73 percent is made of dark energy: an unknown essence that has caused the Hubble expansion of the universe to speed up. In short, the universe is a puzzle for which practically all of the pieces are missing. Could the LHC help track these pieces down?

Anticipation of the missing matter dilemma dates back to long before the issue gained wide acceptance. The first inkling that visible material couldn't be the only hand pulling on the reins of the universe came in 1932, when Dutch astronomer Jan Oort found that stars in the outer reaches of our galaxy moved in a way consistent with much greater gravitational attraction than observed matter could exert. The Milky Way is in some ways like a colossal merry-go-round. Stars revolve around the galactic center and bob up and down relative to the galactic disk. Oort found that he could measure these motions and calculate how much total gravitational force the Milky Way would need to exert to tug stars back toward its disk and prevent them from escaping. From this required force, he estimated the Milky Way's total mass, which became known as the Oort Limit. He was surprised to find that it was more than double the observed mass due to shining stars.

The following year, Bulgarian-born physicist Fritz Zwicky, working at Caltech, completed an independent investigation of the gravitational "glue" needed to keep a massive group of galaxies called the Coma Cluster from drifting apart. Because the galaxies within this formation are widely separated, Zwicky estimated a very high figure for the gravity needed. Calculating the amount of mass needed to furnish such a large force, he was astounded to discover that it was hundreds of times that of the luminous matter. Some invisible scaffolding seemed to be providing the support required to hold such a far-flung structure together.

Scientists in the 1930s knew little about the cosmos, aside from Hubble's discovery of its expansion. Even the concept of galaxies as "island universes" beyond the Milky Way was relatively new. With physical cosmology in such an early stage of development, the astonishing findings of Oort and Zwicky were largely ignored. Decades passed before astronomers acknowledged their importance.

We owe current interest in dark matter to a courageous young astronomer, Vera Cooper Rubin, who entered the field when women were often discouraged from pursuing it. Rubin was born in Washington, D.C., and her childhood hobbies included stargazing from her bedroom window and reading astronomy books—particularly a biography of comet discoverer Maria Mitchell. Much to her frustration, the vocation then seemed a clubhouse with a No Girls Allowed sign prominently displayed.

As Rubin later recalled: "When I was in school, I was continually told to go find something else to study, or told I wouldn't get a job as an astronomer. But I just didn't listen. If it's something you really want to do, you just have to do it—and maybe have the courage to do it a little differently."[1]

After earning a B.A. at Vassar, where Mitchell once taught, and an M.A. at Cornell, Rubin returned to her native city to pursue graduate studies in astronomy at Georgetown University. Though not on Georgetown's faculty, George Gamow, with whom she shared an interest in the behavior of galaxies, was permitted to serve as her thesis adviser. Under his valuable supervision, she received her Ph.D. in 1954.

While raising four children with her husband, mathematician Robert Rubin, it took some time for her to find a permanent position that offered her suitable flexibility. In 1965, the Department of Terrestrial Magnetism of Carnegie Institution in Washington appointed her to its research staff. She soon teamed up with a colleague, Kent Ford, who had built his own telescope. Together they began an extensive study of the outer reaches of galaxies.

Focusing on the Milky Way's nearest spiral neighbor, the Andromeda galaxy, Rubin and Ford used a spectrograph to record the Doppler shifts of stars on its periphery. A Doppler shift is an increase (or decrease) in frequency of something moving toward (or away) from an observer. The amount of the shift depends on the moving object's relative velocity. It occurs for all kinds of wave phenomena—light as well as sound. We notice the Doppler effect when fire engines wail at higher and higher pitch when racing closer, and lower and lower when speeding away. For light, moving closer means a shift toward the bluer end of the spectrum (a blue shift, for short), and moving away, a shift toward the redder end (a red shift). Hubble used galactic red shifts in his proof that remote galaxies are receding from us. Doppler spectroscopy has continued to serve as a vital tool in astronomy.

Mapping out the shifts in light spectra of Andromeda's outermost stars, Rubin and Ford were able to calculate their velocities. They determined how quickly these outliers orbited the galaxy's center. Plotting stars' orbital speeds versus their radial distances, the Carnegie researchers produced an impressive graph, called a galactic rotation curve, displaying how Andromeda steered its remotest material.

As Kepler discovered centuries ago, for astronomical situations, such as the solar system, for which the bulk of the material is in the center, objects take much longer to orbit the farther they are from the middle. The outer planets orbit much slower than the inner ones. While Neptune orbits at a tortoiselike 3.4 miles per second, Mercury whizzes around the Sun at an average pace of 30 miles per second. The reason is that the gravitational influence of the Sun drops off sharply at large radial distances and there is not enough mass in the outer solar system to affect planetary speeds to a large degree.

Spiral galaxies such as the Milky Way were once thought to have a similar central concentration of material. Visibly, the densest concentration of stars lies in bulges around their middles. The outer spiral arms and haloes surrounding the main disks seem, in contrast, to be wispy and dilute. But appearances can deceive.

The Carnegie researchers connected the dots on Andromeda's rotation curve. Fully expecting to see the velocities drop off with radial distances, as in the solar system, they were baffled when their points, even in the outermost reaches, continued along a flat line. Rather than a mountain slope, the curve resembled a level plateau. A flat-velocity profile meant a sprawling out in mass beyond the frontiers of the observed. Something unseen was lending gravitational strength to places where gravity should be puny.

To see if their results were peculiar to Andromeda or more general, Rubin and Ford teamed up with two of their colleagues, Norbert Thonnard and David Burstein, to investigate sixty additional spiral galaxies. Although not all galaxies are spiral—some are elliptical and others are irregular—they chose that pinwheel-like shape because of its simplicity. Unlike other galactic types, the outer stars in spirals generally revolve in the same direction. For that reason, their speeds are easier to plot and analyze.

Relying on data collected by telescopes at Kitt Peak Observatory in Arizona and Cerro Tololo Observatory in Chile, the team members plotted out rotation curves for all sixty galaxies. Amazingly, each exhibited the same leveling off in velocities observed for Andromeda. Rubin and her coworkers concluded that most of the material in spirals is spread out and invisible—revealing nothing about its content except its weighty influence. The mystery that had so troubled Oort and Zwicky was back in full force!

What lies behind the mask? Could dark matter be something ordinary that's simply very hard to see? Could our telescopes just not be powerful enough to reveal the bulk of material in space?

A one-time leading dark matter candidate carries a name that matches its supposed gravitational brawn: MACHOs (Massive Compact Halo Objects). These are massive bodies in the haloes of galaxies that radiate very little. Examples include large planets (the size of Jupiter or greater), brown dwarfs (stars that never ignited), red dwarfs (weakly shining stars), neutron stars (collapsed stellar cores composed of nuclear material), and black holes. Each of these was formed from baryonic matter: the stuff of atomic nuclei and the like, such as hydrogen gas.

To hunt for MACHOs and other hard-to-see gravitating objects, astronomers developed a powerful technique called gravitational microlensing. A gravitational lens is a massive object that bends light like a prism. It relies on Einstein's general relativity theory that heavy objects curve space-time, which in turn distorts the paths of light rays in their vicinity. This was verified in the 1919 observations of the bending of starlight by the Sun during a solar eclipse.

Microlensing is a way of using the gravitational distortion of light to weigh potential MACHOs when they pass between distant stars and Earth. If an unseen MACHO happened to move in front of a visible star (from a neighboring galaxy in the background, for instance), the starlight would brighten due to the MACHO's gravitational focusing. After the MACHO moves on, the light would dim again, back to its original intensity. From this brightness curve, astronomers could determine the MACHO's mass.

During the 1990s, the MACHO Project, an international group of astronomers based at Mt. Stromlo Observatory in Australia, catalogued thirteen to seventeen candidate microlensing events. The team discovered these characteristic brightness variations during an extensive search of the galactic halo using the Large Magellanic Cloud (a smaller neighboring galaxy) to provide the stellar background. From their data, the astronomers estimated that 20 percent of the matter in the galactic halo is due to MACHOs, ranging from 15 percent to 90 percent of the mass of the Sun. These results point to a population of lighter, dimmer stars in the Milky Way's periphery that cannot directly be seen but only weighed. Though these objects might add some heft to the galactic suburbs, the MACHO Project has shown that they could account for only a fraction of the missing mass.

There are other reasons to believe that MACHOs could help resolve only part of the dark matter mystery. Using nucleosynthesis (element-building) models that estimate how many protons must have been present in the moments after the Big Bang to produce the elements we see today, astrophysicists have been able to estimate the percentage of baryonic matter in the universe. Unfortunately, these estimates show that only a small fraction of dark matter could be baryonic in nature; the rest must be something else. Made of conventional baryonic matter, MACHOs thereby could not provide the full explanation. Consequently, researchers have turned to other candidates.

The beefy acronym MACHO was chosen to contrast it with another class of dark-matter candidates, the ethereal WIMPs (Weakly Interacting Massive Particles). Unlike MACHOs, WIMPs would not be astronomical objects but rather new types of massive particles that interact exclusively through weak and gravitational forces. Because of their heaviness, they'd be slow moving—enabling their gravitational "glue" to help cement together the large structures in space we observe, such as galaxies and clusters.

Neutrinos would fit the bill if they were heavier and more lethargic, given that as leptons, they ignore the strong force, and as neutral particles, they pay no heed to electromagnetism. The lightness and swiftness of neutrinos, however, seems to rule them out as significant components. This fleeting nature is akin to a featherweight, constantly traveling politician trying to draw support for a local council race. Without setting down robust roots in his community, how could he bring people together? Similarly, neutrinos never hang around long enough or make enough of an impact to serve as uniters.

Particles such as neutrinos that would be too light and quick to create structure are sometimes referred to as "hot dark matter." Although they might compose a portion of the missing mass in the universe, they could not explain how galaxies maintain such tight holds on their outermost stars nor how they clump together into clusters. Slower, bulkier substances, such as

MACHOs and WIMPs, are grouped together into "cold dark matter." These would offer suitable scaffolding—if we could only find enough of them.

If not neutrinos, then which other neutral, nonhadronic particles could carry enough mass and move at a slow enough pace to steer stars and gravitate galaxies? Unfortunately, the standard model doesn't call any suitable candidates to mind. Apart from neutrinos, MACHOs, and WIMPs, another option, a hypothetical massive particle called the axion, postulated to play a role in quantum chromodynamics (the theory of the strong force) and tagged by some theorists as a leading dark-matter contender, has yet to be found. The search for the universe's missing mass has been at an impasse.

Enter the LHC to the rescue. Perhaps somewhere in its collision debris the secret key ingredients of cold dark matter will be revealed. Prime contenders would be the lightest supersymmetric companion particles, such as neutralinos, charginos, gluinos, photinos, squarks, and sleptons. Presuming they have energies on the TeV scale, each would present itself through characteristic decay profiles that would show up in tracking and calorimetry.

If dark matter were the main cosmic mystery, physicists would simply be clenching their teeth, crossing their fingers, and waiting expectantly for results at the LHC or elsewhere to turn up a suitable prospect. It would be like posting a reasonable job description and hoping that the right person will eventually apply. However, a much more nebulous search—the quest for dark energy—has turned out to be far more unnerving. Not only is something seriously missing, but scientists have little idea where to look.

Dark energy first jolted the scientific community in 1998, when two teams of astronomers—a group from Lawrence Berkeley National Laboratory led by Saul Perlmutter and a collaboration based at Mt. Stromlo Observatory that included Adam Riess, Robert Kirschner, and Brian Schmidt—announced startling results about the expansion of the universe. Each used supernovas in remote galaxies as distance gauges to trace the cosmic expansion far back in time. By plotting the distances to these galaxies versus their velocities as found by Doppler redshifts in their spectral lines, the teams could determine how Hubble's law of galactic recession has changed over billions of years.

The type of exploding stars examined, called Supernova Ia, has the special property that their energy produced follows a regular progression. Because of this predictability, the teams were able to compare their actual with their observed light outputs and calculate how far away they are. This offered a yardstick to galaxies billions of light years away—recording their distances at the time of the stellar burst.

Astronomical objects with known energy output are called standard candles. Like distant street lamps on a dark road, you can judge their remoteness by how bright or dim they seem—assuming that they put out roughly the same wattage. If, when walking down a street at night, your eyes were dazzled by an intense glare, you would likely deem its light source much closer than if it were so faint that you could barely see it. You would thereby be able to use its relative brightness to estimate its distance. Similarly, astronomers rely on standard candles such as Supernova Ia to gauge distances for which there would be no other measure.

The team led by Perlmutter, called the Supernova Cosmology Project (SCP), has deep connections with the world of particle physics. First of all, along with George Smoot's Nobel Prize–winning exploration of the cosmic microwave background using the Cosmic Background Explorer satellite, it represents an expansion of the mission of Lawrence's lab. Given that Lawrence was always looking for connections and applications, such a broad perspective perfectly suits the former Rad Lab. Also, one of the SCP's founding members is Gerson Goldhaber, who won acclaim for his role in the Stanford Linear Accelerator Center–led group that jointly discovered the J/psi particle. His older brother Maurice Goldhaber worked at Cavendish during the

Rutherford/Chadwick era and was the longtime director of Brookhaven National Laboratory. So you could say that cosmology and high energy physics—the sciences of the extremely large and extraordinarily small—have become part of the same family.

When the SCP began its explorations, its researchers hoped to use supernova standard candles as means of pinning down the *deceleration* of the universe. The attractive nature of gravity means that any set of massive objects moving apart must reduce its outward rate of expansion over time. Simply put, what goes up must come down—or at least slow down. Cosmologists therefore expected that the dynamics of the cosmos would follow one of three different paths, depending on the universe's density relative to a critical value: brake rapidly enough to reverse course, brake gradually enough not to reverse course, or brake at the precise rate required to remain perpetually on the cusp.

All three scenarios would begin with the standard Big Bang. If the density were high enough, the universe would slow down enough over time that after many billions of years its expansion would transform into a contraction. Eventually, everything would compress back together in a Big Crunch. If the density were lower than the critical value, on the other hand, the cosmic expansion would forever continue to slow down—like a tired runner sluggishly pushing himself forward. Though galaxies would move apart at an increasingly lethargic pace, they'd never muster the will to reunite. This possibility is called the Big Whimper. A third option—the density exactly matching the critical value—would involve a universe slowing down so much that it threatened recollapse but never quite did, like an acrobat carefully poised on a tightrope.

Perlmutter and his team fully expected to encounter one of these three possibilities. Their supernova data surprised them with a different story. Plots of velocity versus distance showed that the cosmic rate of expansion was speeding up, not slowing down. Something was pressing the gas pedal instead of the brakes—and it couldn't be any of the known forces. University of

Chicago theorist Michael Turner dubbed this unknown agent "dark energy."

While both have mysterious identities, dark energy could not be the same as dark matter. In contrast to dark matter, which would gravitate in the same way as ordinary matter, dark energy would serve as a kind of "antigravity," causing outward acceleration. If dark matter walked into a party, it would serve as a graceful host introducing people and bringing them together, but if dark energy intruded, it would act like the riot police dispersing the crowd. Indeed, too much dark energy in the cosmos would be no fun at all—the universe would eventually tear itself apart in a catastrophic scenario called the Big Rip.

Some physicists have represented dark energy by restoring Einstein's once-discarded cosmological constant term to general relativity. Although adding such a constant antigravity term would be a simple move, it could use some physical motivation. Physicists would be loath to add anything to a well-established theory without understanding the need for the new term on a fundamental level. That would mean interpreting the field theory behind it. Current field theories, however, support a much larger value of the vacuum energy that would need to be almost, but not exactly, canceled out to yield a reasonable cosmological constant. Thus, matching experimental bounds for cosmic acceleration has proven a daunting task.

Moreover, if dark energy were a constant throughout space and time, it would never lose its effect. With gravity ceding more and more ground over the eons to dark energy, the Big Rip would be an absolute certainty. Before accepting such an outcome as inevitable, most theorists would like to mull over the alternatives.

Princeton physicist Paul Steinhardt, along with theorists Robert Caldwell and Rahul Dave, has suggested a different way of modeling dark energy, through a wholly new type of substance called quintessence. Quintessence is a hypothetical material with negative pressure that pushes things apart (like an elemental Samson on the Philistines' columns) rather than pulling them together (like ordinary, gravitating matter). Its name harks back to the four classical elements of Empedocles—with quintessence representing the fifth. The distinction between a cosmological constant and quintessence is that while the former would be as stable as granite, the latter could vary from place to place and time to time like moldable putty.

Findings of the Wilkinson Microwave Anisotropy Probe of the cosmic microwave background support the idea that the cosmos is a mixture of dark energy, dark matter, and visible matter—in that order. The satellite picture has not been able to tell us, however, what specific ingredients constitute the duet of dark substances.

Physicists hope that further clues as to the nature of dark energy, as well as dark matter, will turn up at the LHC. The discovery of quintessence at the LHC, for example, would revolutionize the field of cosmology and transform our understanding of matter, energy, and the universe. Indeed, based upon what is found, the fate of everything in space could be in the balance.

Adding a cosmological constant or postulating a novel kind of material are not the only alternatives. Some theorists see a need to rethink the nature of gravity completely. Could gravitation behave distinctly on different scales—acting one way in the planetary arena and another in the galactic realm? Might Einstein's equations of general relativity, accurate as far as we can judge, be superseded in the grandest domain by another theory? As Rubin has said, "I suspect we won't know what dark matter is until we know what gravity is."[2]

Radical new gravitational theories propose a fundamental change in its mechanism and scope. They imagine that some of gravity's properties could be explained through its ability to penetrate unseen extra dimensions impervious to other forms of matter and energy. Conceivably, the dark substances in the universe could be shadows of a higher reality.

Remarkably, some of these novel theories, as strange as they seem, could be tested at the LHC. The extraordinary power of high-energy transformations may well reveal new dimensions in addition to novel particles. Who knows which of nature's long-held secrets the LHC's unprecedented energies will divulge?

10

The Brane Drain

Looking for Portals to Higher Dimensions

If this is the best of all possible worlds,
what are the others like?

—VOLTAIRE, *CANDIDE* (1759)

The LHC, with its matter-changing properties, could be said to offer a modern-day "philosopher's stone." Interestingly, its region's stone has already been used to build a philosopher's house. That philosopher was François-Marie Arouet, better known as Voltaire.

The Chateau de Ferney, where the witty writer lived from 1758 almost until his death in 1778, is situated within a mile of the ring traced by the LHC. Within that mansion he completed his most famous work, *Candide*, a cutting satire of the optimism of the German thinker Gottfried Leibniz. At first glance, the LHC and Leibniz (and Voltaire's parody thereof) might seem to have little in common. However, they are profoundly connected through the concept of parallel universes and alternative realities.

A rival of Newton in developing calculus, Leibniz believed that our world represents the optimum among the set of all possibilities. Leibniz drew his conclusions from the calculus of variations—the method he developed for finding shortest paths along a surface and related problems. An example of such a situation considers the ideal way to cross a hill—among the myriad routes there is one that minimizes the length. Leibniz pondered that God, in designing the universe, would choose the optimal solution whenever there are alternatives.

Voltaire's farcical character Dr. Pangloss, a teacher of "metaphysico-theologico-cosmolonigology," takes this concept to the extreme by concocting a convoluted rationale for anything that happens in this "best of all possible worlds."

"Observe that noses were made to support spectacles," remarks Pangloss. "Hence we have spectacles. Legs were obviously instituted to be clad in breeches, and we have breeches."[1]

Even after Pangloss, along with his pupil Candide, suffers through the most horrific series of events imaginable, including the destruction wrought by the Lisbon earthquake and the terror of the Inquisition, he continues to rationalize his experiences. He concludes that if a solitary link were broken in the cosmic chain of events, no matter how dreadful the occurrence seemed at the time, ultimate good would never ensue—in their case, the possibility to cultivate a small garden. No one following the dismal adventure could miss Voltaire's irony.

Could we be living in the "best of all possible worlds?" The concept implies the existence of alternative realities—perhaps even universes parallel to our own. Until recently the concept of parallel universes lay exclusively in the realm of speculation. However, remarkably, one of the projects planned for the LHC is to test a new type of parallel universe idea, called the "braneworld" hypothesis, positing that everything we observe resides within a three-dimensional island surrounded by a sea of higher dimensions. "Brane" is short for "membrane," a description of structures such as the one theorized to support the observable cosmos. According to this hypothesis, the only particles able to leave our brane are gravitons, which are the carriers of gravity. Consequently, researchers plan to use the LHC to search for gravitons leaking into higher dimensions. If such extra dimensions are found, perhaps there are other branes parallel to our own. If these other branes turn out to be lifeless structures, perhaps we do indeed live in the optimum world.

The concept of parallel universes first entered physics in a kind of abstract, mathematical way through Richard Feynman's diagrammatic method of calculating the likelihoods of certain types of exchanges between charged particles. Each possibility is assigned a certain weight and added up. One way of expressing this "sum over histories" makes use of Leibniz's calculus of variations through what is called the path integral formulation. According to this approach, if you know the starting and ending states of any quantum interaction, what happens in between is like a "hill" of multiple trajectories. You never know exactly which way the players in the interaction breached the hill; in fact, they traversed it many ways at once. All you can calculate is the most probable means of crossing, which works out to be the shortest distance.

Feynman did not intend his method, which he began to develop in the early 1940s under the supervision of his thesis adviser John Wheeler, to represent paths through a labyrinth of actual parallel universes. The math worked out splendidly and the predictions turned out perfectly accurate, that's all. However, in 1957, another of Wheeler's students, Hugh Everett, took matters a step further through his "Many Worlds Interpretation" of quantum mechanics.

According to Everett's hypothesis, each time particles interact on the microscopic level, the universe responds by bifurcating into a maze of alternatives, each slightly different. When an experimenter measures the result, he or she replicates into versions corresponding to each alternate reality. Each copy records a different result of the measurement, attributing the outcome to chance. But in reality, there is no chance involved, because every possibility is actually realized by a replica researcher—unable to communicate with other versions and compare results. Over time, the number of parallel universes—and occupants within—grows into a staggering figure, dwarfing even the number of atoms in all of visible space.

Despite this shocking conclusion, in the 1970s noted theorist Bryce DeWitt became convinced of the importance of Everett's conjecture. DeWitt named and popularized the concept, arguing that it was the only reasonable way to make quantum mechanics objective, rather than dependent on the subjective act of measurement. After all, who could step outside of the universe itself, take readings, and cause its wave function to collapse into various possibilities? As crazy as the Many World Interpretation sounds, he argued, isn't it crazier to think of humans influencing the cosmos through their sensory perceptions? Although by then Everett had left theoretical physics (and would die in 1982 at the age of fifty-one), DeWitt was very effective in promoting the idea that we live in an ever-expanding web of parallel universes.

Along with Wheeler, DeWitt had already made inroads into the question of applying quantum principles to gravity. Wheeler was very interested in developing a sum-over-histories method for encapsulating the solutions of Einsteinian general relativity. In quantum mechanics, states can have different positions, momenta, spins, and so forth. These are like the distinct musical notes that make up a composition. What would be the equivalent keyboard for general relativity? Eventually, it occurred to him that the range of possible three-dimensional geometries would offer the medley of tones needed to compose his symphony. Exuberantly, he prodded DeWitt to help him develop the mathematical notation for this idea. As DeWitt recalled:

> Wheeler used to bug everybody. I got a telephone call from him one day around 1964 saying he would be passing through the Raleigh-Durham airport—that's

> when I was in North Carolina—between planes for two hours. Would I please come out there and we would discuss physics? I knew that he was bugging everybody with the question "What is the domain space for quantum gravity?" And I guess he had it finally figured out in his mind that it was the space of three-geometries. This was not the direction I was really concentrating my efforts, but it was an interesting problem so . . . I wrote down this equation. I just found a piece of paper out there in the airport. Wheeler got very excited about this.[2]

The result was the Wheeler-DeWitt equation: a way of assigning weights to three-dimensional geometries and summing them up to determine the most probable evolution of the universe. In theory, it was supposed to help researchers understand how reality as we know it emerged from the chaotic jumble of possibilities. In practice, however, the equation would become unwieldy if applied to complex situations.

In 1973, C. B. Collins and Stephen Hawking considered this question classically in their influential paper "Why Is the Universe Isotropic?" Pondering the myriad possible general relativistic solutions—including isotropic as well as anisotropic cosmologies—they wondered which could evolve into the familiar present-day universe. The difference between isotropic and anisotropic cosmologies is that while the former expands evenly in all directions, like a spherical balloon being filled with air, the latter blows up at unequal rates depending on which way you look, more like a hotdog-shaped balloon becoming longer and longer as it is inflated but not much wider.

Not surprisingly, according to astronomers, the present-day universe on the largest scales is close to isotropic. Space seems to be expanding close to the same rate in all directions. The cosmic microwave background, a snapshot of the "era of recombination" three hundred thousand years after the birth of the universe, is similarly very close to being isotropic. (As we discussed, the COBE [Cosmic Background Explorer] and WMAP

[Wilkinson Microwave Anisotropy Probe] satellites mapped out minute anisotropies.) Collins and Hawking wondered whether the very early universe, instants after the Big Bang, needed to have been isotropic as well. Why couldn't it have been arbitrarily chaotic like the hodgepodge of sand dunes on a rugged beach?

To examine the possibility of cosmic evolution from chaos to order, they considered what is now called the multiverse: a kind of universe of universes embodying the range of all geometric possibilities. Of this cosmic zoo, they wondered, which kind of creatures could evolve into the tame entity with which we are familiar: isotropic space as we see it today. Surprisingly, according to their calculations, only an infinitesimally minute percentage could make the leap. Only universes that were extraordinarily isotropic to begin with could end up with ordinary present-day conditions. Any deviation from perfection in the beginning would blow up over time into a cosmic monstrosity. How then to justify the improbable normality of today?

In lieu of an explanation based exclusively on physical laws, Collins and Hawking decided to invoke what Australian physicist Brandon Carter dubbed the anthropic principle: the concept that the existence of humans constrains the nature of the universe. If the universe were sufficiently different, anthropic reasoning asserts, stars like the Sun wouldn't have formed, planets like Earth would be absent, beings like humans would not exist, and there would be nobody to experience reality. Therefore the fact that we, as intelligent entities, are around implies that the universe must have been close enough to its present form to guarantee the emergence of such cognizant observers. Collins and Hawking applied the anthropic principle as follows to explain why the universe is isotropic:

> Suppose there are an infinite number of universes with all possible different initial conditions. Only those universes which are expanding just fast enough to avoid recollapsing would contain galaxies, and hence intelligent life.

> [These] would in general approach isotropy. On this view, the fact that we observe the universe to be isotropic would be simply a reflection of our own existence.[3]

The use of anthropic reasoning is akin to compiling news clippings about lottery results around the world and realizing that the reason for all of the success stories is that coverage is biased in favor of winners. Although there are millions of "parallel histories" of people who buy lottery tickets, only those who hit the jackpot generally make the news. If you perused all lottery stories without knowing this, you might wonder if lotteries almost always pay off handsomely. Not only would this seem unprofitable for those running such contests, it would also appear to violate the laws of chance. However, the selection principle of newsworthiness strongly favors the minute subset of parallel histories that end up in success. Similarly, the selection principle of conscious observation strongly favors the minute subset of parallel universes that end up producing intelligent life.

Throughout the final decades of the twentieth century—with respected physicists such as DeWitt, Collins, and Hawking referring in their research to a large or even infinite tally of universes—the speculative concept of alternative realities became a serious scientific talking point. Theorists grew bolder in their allusions to parallel realms beyond the reach of telescopic surveys. If a physical parameter couldn't be nailed down through an analysis of the observed universe, many researchers began to rely on a toolbox of effects based on the supposition of a largely unseen multiverse.

In 1980, American physicist Alan Guth proposed cosmic inflation as a potential solution of a number of issues in modern cosmology, including the question of why the present-day universe is so uniform. Instead of invoking anthropic reasoning, he suggested that the very early universe went through a stage of ultrarapid expansion that stretched out all abnormalities beyond the point of observability—similar to pulling on a bed sheet to smooth out the wrinkles. Guth's initial theory, though promising, presented a number of quandaries, including a prediction of observable transition zones between sectors of the universe possessing different conditions. Because astronomy doesn't record such barriers, the theory required modification.

Three years later, Russian cosmologist Andrei Linde linked the inflationary concept with the multiverse idea through a novel proposal called chaotic inflation. In Linde's variation, the multiverse is a nursery harboring the seeds of myriad baby universes. These seeds are sown through a randomly fluctuating scalar field (something like the Higgs but more variable) that sets the value of the vacuum energy for each region. Through the general relativistic principle that mass and energy govern geometry, the places where this energy is highest stimulate the fastest-growing areas—such as the abundance of jobs triggering growth in certain communities. As in sprawling suburbs plowing over fallow farms, the most rapidly expanding parts of the universe—the inflationary regions—quickly dominate all of the others. Linde's conclusion was that we live in one of these hyperexpanded megalopolises—with any others long since nudged away beyond possible detection.

Inflation has become a popular way of understanding the overall uniformity of the observable cosmos. One of its key advantages over pure anthropic argumentation is that it doesn't rely on the existence of humans to explain how tapioca pudding–like blandness emerged from the bubbling chaos of the primordial universe. Yet, by literally pushing alternative versions of our universe beyond measurement, inflationary cosmology removes a potential means of verification. Fortunately, it offers predictions about the distribution of matter and energy in the stages of the universe after inflation. These characteristic patterns manifest themselves in the cosmic background radiation, which has been analyzed by WMAP and other surveys. The consensus of astrophysicists today is that cosmic inflation in some form remains a viable explanation of how the early universe developed. What form of inflation might have occurred and what could have caused such an era remain open questions.

The latest breed of parallel universe theory, the braneworld hypothesis, relies not on unseen realms of our own space but rather on dimensions beyond the familiar three. The far-reaching idea proposes that ordinary space comprises a three-dimensional membrane—or "brane," for short—floating in a higher dimensional reality called the bulk. According to this notion, the bulk would be impervious to all particles except gravitons. Because the carriers of the electroweak and strong interactions cannot penetrate its depths, its existence would affect only gravitational interactions. Hence, without photons being able to enter the bulk, we could not see it. The dilution of gravity by means of gravitons leaving our brane and infusing the bulk would explain why gravity is so much weaker than the other forces.

The concept of branes is a variation of string theory that generalizes the jump rope–like vibrations of string into pulsating objects of two, three, or higher dimensions—akin to bouncy trampolines or shimmering raindrops. These could have an enormous span of sizes—ranging from minute enough to represent elementary particles to grand enough to encompass all of observable space. From the latter stems the idea that everything around us, except for gravitons, lives on a brane.

Branes have been under discussion as particle models for several decades. Dirac conceived the idea in the 1960s that particles are extended rather than pointlike. He didn't push the concept very far, however, and it was scarcely noticed by the physics community. In 1986, Texas researchers James Hughes, Jun Liu, and Joseph Polchinski constructed the first supersymmetric theory of membranes, demonstrating how they could represent various types of particles. The following year, Cambridge physicist Paul Townshend coined the term *p-branes* to denote higher-dimensional extended objects dwelling in an eleven-dimensional reality—like curiously shaped peas living in an exceptionally spacious and intricate pod. (The "p" takes on values representing the number of dimensions of the membrane.)

Around the same time, Townshend, his colleague Michael Duff, and other theorists revealed deep connections between membranes and strings called dualities. A duality is a kind of mathematical equivalence that allows swaps between the extreme cases of certain variables—for example, exchanging a microscopically small radius for an enormous one—while preserving other physical properties. It is like a card game in which the numbers "1" and "11" are both represented by aces, allowing players with aces to switch their value strategically from low to high to wield the best hand. Similarly, there are cases in membrane theory in which flipping certain variables from small to large serve well in proving certain equivalences.

Membrane theory was little noticed by the mainstream physics community until the mid-1990s, when a combination of dualities developed by researchers in that area served to unite the five kinds of string theory. When string theory first came into prominence in the early 1980s as a potential "theory of everything," various theorists proposed an embarrassing assortment of types—technically known as Type I, Type IIA, Type IIB, Heterotic-O, and Heterotic-E—each of which seemed suitable. How to distinguish which was the real deal? Surely a theory of everything must be unique.

It would be like different witnesses to a crime scene relating clashing descriptions to a detective—with one saying, "He had a long gray coat," another indicating, "He was wearing a short blue vest," and so forth—until the sleuth figured out that shadows and lighting altered the culprit's appearance. An awning blocking the Sun from a certain angle darkened his jacket and made it seem longer. Similarly, dualities gleaned from membrane theory showed that by changing perspectives all five varieties of string theory can be transformed into one another.

At a 1995 conference in Southern California, the leading string theorist Ed Witten dramatically announced the discovery of the "duality of dualities" uniting all of the string theory brands into a single approach, which he called "M-theory." Rather than defining the term, he left its meaning open to interpretation, asserting that the "M" could stand for "magic," "mystery," or "matrix." Others thought immediately of "membranes" and "mother of all theories." The excitement generated by that announcement and the realization that string theory could be unified heralded what became known as the second string revolution (the first revolution being the 1980s discovery that string theory doesn't have mathematical anomalies).

In the unification of string theory, one of the parameters found to be adjustable is the size of what is called the large extra dimension. This nomenclature distinguishes several distinct kinds of dimensions. First of all, there are the three dimensions of space, length, width, and height, which, along with the dimension of time, make up four-dimensional space-time. Second, following an approach first suggested by Swedish physicist Oskar Klein, there are the small "compactified" dimensions—those curled up into tight knots too minuscule ever to observe. According to the ideas of Witten and others, these form various types of six-dimensional clusters named Calabi-Yau spaces after mathematicians Eugenio Calabi and Shing-Tung Yau. Finally, there is an eleventh dimension of adjustable size—with dualities enabling it to thicken like dough mixed with yeast. This large extra dimension could conceivably be of detectable proportions.

How can we envision an extra dimension perpendicular to those we normally experience? It's like describing a hot air balloon ride to people who have never left the ground. Before the age of ballooning, nobody ever experienced Earth from an aerial perspective. Balloons—and later airplanes and spaceships—permitted far greater exploration of the dimension of height. If the eleventh dimension exists, and it is not curled up, what prevents us from experiencing that, too? According to some theorists, the answer may lie in the stickiness of the strings that make up matter and luminous energy.

A critical aspect of M-theory is the concept of the Dirichlet brane, or "D-brane" for short, developed by UC Santa Barbara researcher Joseph Polchinski, along with J. Dai and R. G. Leigh. Polchinski defined these as extended objects to which the end points of open strings can be attached. Open strings are those not connected with themselves, rather hanging loose like strands of spaghetti. The opposite of these are closed strings, which form complete loops like onion rings. Polchinski and his colleagues showed that open strings naturally cling to D-branes as if their ends were made of glue, but closed strings have no such constraint.

String theory represents quarks, leptons, photons, and most other particles as open strings. The exception is gravitons, modeled by closed strings. Therefore, aside from gravitons, all particles would naturally stick to a D-brane. Gravitons, on the other hand, would be free to wander away from one D-brane and head, like migrating birds, toward another.

The dichotomy between the stringy behavior of gravitons and other particles suggested a way to model the relative weakness of gravity using M-theory and resolve the hierarchy problem described earlier. In 1998, Stanford physicists Nima Arkani-Hamed, Savas Dimopoulos, and Gia Dvali (joined for one paper by Ignatius Antoniadas) suggested a scenario called the ADD model (after their initials) involving two D-branes separated by a large extra dimension on the order of 1⁄25 of an inch in size. The second D-brane would represent a parallel universe, or another section of our own universe, right in front of our eyes but completely invisible. Because all standard model fields would remain confined to our own brane, photons would never be able to make the leap and illuminate the parallel brane. The strong and weak forces would similarly keep mum about the close but hidden realm. Instead, the only means of discerning its existence would be through the unseen tugs of gravity.

Due to its ability to fill the bulk in between our brane and the parallel brane, gravity would become diluted—making it much less powerful than the other interactions. It would be similar to four boilers in the basement of a ten-story apartment building, with the first three used to provide steam for a spa and sauna in an adjacent room, and the fourth pumping out heat for the other floors. While those in the spa might enjoy the full force of the steam, those on the highest floor might be huddled under blankets in chilly rooms. The strength of the boilers might be the same, but the dilution of the steam the fourth one produced would make it much less effective. Similarly, the leakage of gravitons from our brane would allow them to have the same interactive strength in principle as other exchange particles—weakened only because of their seepage into the bulk.

Unlike the MSSM (Minimal Supersymmetric Standard Model) method of augmenting the Standard Model with supersymmetry, the large extra dimensions approach offers the advantage of a single unification energy rather than several. Everything would be united on the TeV scale, which is, conveniently, the energy of the LHC. Gravity would just appear weaker because of its secret excursions to a nether realm. The tremendous energy of the Planck scale, which is much higher than the TeV scale, would never have to be approached to test unification. If that's true, it would sure save a lot of money.

Large extra dimensions also offer the enticing possibility of helping astronomers understand dark matter. In a variation of the scheme, called the Manyfold Universe, the ADD team, along with Stanford physicist Nemanja Kaloper, pondered what would happen if our brane were folded up like an accordion. Stars that are distant along our brane—their light taking millions of years to reach us—could be close together by way of a shortcut through the eleventh dimension. It would be the same as standing at the end of a serpentine line winding around a zigzag chain-link fence and then having someone lift a chain that allows you suddenly to be right next to the person formerly well ahead of you.

Because gravitons could breach the shortcut through the bulk, two otherwise distant stars could gravitationally influence each other. This influence would be felt but not seen, offering a possible explanation for at least one form of dark matter. In other words, some types of dark matter would actually be luminous matter pulling on other luminous matter through the curtain of the bulk.

One of the appealing aspects of the ADD scenario is that it offers testable predictions. It presages modifications to the law of gravity on scales less than 1⁄25 of an inch. In contrast to the conclusions of both Newton and Einstein, it hypothesizes that gravitational attraction, for those short distances, would no longer follow an inverse-squared behavior but would rather be modified by an additional factor. For large distances, such as the radius of the moon's orbit around the Earth, this factor would be insignificant, explaining why the discrepancy has never been noticed. Unfortunately for the success of the theory, despite numerous tests using sensitive instruments, the discrepancy has yet to be detected on the smallest scales, either. For example, experiments by a group led by Eric Adelberger at the University of Washington using a delicate kind of twisted pendulum called a torsion balance have measured gravitational attraction to be an inverse-squared relationship for distances much smaller than 1⁄25 inch. This has cast doubt upon at least the simplest form of the theory.

In 1999, physicists Lisa Randall and Raman Sundrum proposed a different kind of braneworld scheme that doesn't require the same stark modifications of the law of gravity. Although, like the ADD model, the Randall-Sundrum model posits two three-dimensional branes—one representing the observable universe where the Standard Model lives; and the other, a kind of forbidden zone where only gravitons dare venture—it doesn't require the distance between the branes to be measurably large. Rather, the parallel branes could be so close together to elude even the most sensitive instruments.

To achieve this feat, Randall and Sundrum found a clever way to dilute gravity without the need for a large extradimensional arena. They proposed a warping of the bulk that would concentrate the greatest part of the gravitons' wave function away from our brane. This warping would be a function of the distance from our brane in the extra dimension—growing deeper like the ocean away from the shore. Consequently, gravitons would have a much higher probability of being in the region near the other brane than touching ours. They would have minimal interaction with the particles on our brane—rendering gravity much weaker than the other forces.

We can envision the distinction between the ADD and Randall-Sundrum models in terms of choices an urban planner might make about accommodations for parking to keep cars away from the main street of a town. One option, analogous to the ADD approach, would be a spacious, flat parking lot nearby. Most cars would be scattered around its interior, far from the street. If there isn't much space to spare, however, the planner might choose instead to dig deep and build an underground garage. Cars entering the garage would follow a ramp downward. As in the flat case, the cars would be well off the street—with the advantage of minimizing the impact on the cityscape. That second option would be more akin to the Randall-Sundrum approach.

To complete the analogy, we can think of Main Street as corresponding to the three-dimensional space in which we live, the number of parked cars on the street as representing the measured strength of gravity, and satellite photos as signifying scientific observation. The situation with no off-street parking opportunities and cars jammed on Main Street would represent gravity as being much stronger in our three-dimensional space than what we actually detect in nature. The case with the flat parking lot would represent weak gravity and a large extra dimension that could easily be spotted by scientists. Finally, the case with the underground garage would similarly allow for weak gravity, but with the bonus of keeping the extra dimension concealed from direct scientific detection. Someone perusing an aerial shot might mistake the community for a quiet town with few cars—like a physicist mistaking our cosmos for one with a paltry number of dimensions and weak gravity.

If the bulk is scooping up gravitons like a plethora of zealous dustpans, might there be a way of detecting the extradimensional leakage with the LHC? One method, already attempted at the Tevatron, would be to look for events in which the detected particles spray in one direction but not another. This imbalance would indicate that an unseen particle (or set of particles) carried away a portion of the momentum and energy. Although this could represent an escaping graviton, more likely possibilities would need to be ruled out, such as the commonplace production of neutrinos. Unfortunately, even a hermetic detector such as ATLAS can't account for the streams of lost neutrinos that pass unhindered through almost everything in nature—except by estimating the missing momentum and assuming it is all being transferred to neutrinos. Some physicists hope that statistical models of neutrino production would eventually prove sharp enough to indicate significant differences between the expected and actual pictures. Such discrepancies could prove that gravitons fled from collisions and ducked into regions beyond.

Another potential means of establishing the existence of extra dimensions would be to look for the hypothetical phenomena called Kaluza-Klein excitations (named for Klein and an earlier unification pioneer, German mathematician Theodor Kaluza). These would reveal themselves as shadows in our brane of particles traveling though the bulk. We'd observe these as particles with the same charge, spin, and other properties as familiar particles except for curiously higher masses.

Plato wrote a famous allegory about prisoners shackled to the interior wall of a cave since childhood, unable to observe the outside world directly. They watch the shadows on the opposite wall and mistake these images for real things. For example, they think that the shadows of individuals carrying vessels as they walk by are actual people. Eventually, one of the prisoners escapes, explores the world outside the cave, and informs the others about their delusion.

Similarly, it is possible that the LHC results (from ATLAS or CMS) could serve as a "cave wall" by which we could observe shadows of particles moving in a greater reality. These particles would have an extra component of their momentum corresponding to their ability to travel along an extra dimension. Because of the unobservability of the extra dimension, we couldn't actually see the particles moving in that direction. Rather, their unseen motion would manifest itself through an additional amount of mass associated with their extra energy and momentum. Researchers hope that the energies of some of the lightest Kaluza-Klein excitations are at the low end of the TeV scale, which could enable them to be observed by LHC researchers.

A host of research articles have offered predictions for potential signals of Kaluza-Klein gravitons and other particles beefed up by extra dimensions. These could decay into electron-positron pairs, muon-antimuon pairs, or other products at energies indicating their possible origin. Studying such excitations would yield valuable information about the size, shape, and other properties of the bulk.

Finding evidence of extra dimensions isn't one of the primary goals of the LHC. However, discovering unseen romping grounds beyond the view stands of our familiar arena would make particle physics a whole new game. Like Plato's cave dwellers, we'd have to face the possibility that everything around us is a shadow of a greater reality. Yet if, on the other hand, visible space plus time make up all that there is, the quest for extra dimensions would ultimately prove futile. Theorists would need to concoct other explanations for why all the other forces are so much more potent than gravity.

One can imagine Voltaire's spirit hovering over the LHC, stirred by the whirlwind of particles circulating beneath his former village of Ferney, and smiling at the search for other possible worlds. Would he have considered it valid science or an exercise in Panglossian "metaphysico-theologico-cosmolonigology"? Perhaps he'd simply be pleased that his haunting grounds remain *un jardin ouvert sur le monde*, cultivated above and below by motivated gardeners to sustain the body and the mind.

11

Microscopic Black Holes

A Boon to Science or a Boom for the World?

Human minds weren't meant to picture something that was smaller than an atom, and yet weighed megatons. . . . Something ineffably but insatiably hungry, and which grew ever hungrier the more it ate.

—DAVID BRIN, *EARTH* (1990)

Now I am become Death, the destroyer of worlds.

— ROBERT OPPENHEIMER, QUOTING THE *BHAGAVAD GITA* FOLLOWING THE TRINITY NUCLEAR TEST

Not all scientists are lunatics. On the contrary, despite cinematic depictions ranging from Dr. Caligari to Dr. Evil, and aside from a smattering of harmless eccentrics, genuinely mad scientists are few and far between. Yet the cultural stereotypes persist, and drive particularly nervous members of the public to assume that the average laboratory researcher would think nothing of taking chances with the fate of the world.

We refer here to civilian scientific research—designed to press forward human knowledge. In an age of consent forms and the omnipresent potential for lawsuits if events go awry, experimental scientists today are generally extremely careful not to expose the public to hazards. People do sometimes make mistakes, but, if anything, scientists are particularly thorough. "Nutty Professor" stereotypes aside, if you read in the paper about a chemical explosion—which would be the more likely cause, an industrial accident or a botched scientific experiment? I would venture to say that it would be the former in almost every case.

It's true that in times of war, scientists have been recruited to conduct far, far riskier experiments. The expectation that war incurs horrific dangers makes that a completely different story. Those involved in the Manhattan Project, for example, knew that they were constructing and testing weapons of unmatched destructive power. In the Trinity test, in which the plutonium bomb was detonated for the first time in the aptly named region of New Mexico called Jornada del Muerto (Journey of the Dead Man), nobody knew for sure what would be the effect of a nuclear explosion. Would the blast be confined to that desert basin or would it race out of control and cause untold damage—perhaps even destroy the world?

In a macabre kind of gambling, right before detonation Fermi offered to take bets on whether a chain reaction would occur that would vaporize the atmosphere. Wagerers could decide if just New Mexico would be wiped out or if it would be the whole Earth. In retrospect, the idea that an experiment could be conducted with an unknown impact on the fate of the entire planet is shocking—and physicists joking about apocalyptic outcomes is highly disconcerting.

As we know now, although the bomb test lit up the sky like a "thousand suns"—as J. Robert Oppenheimer described, borrowing language from the Bhagavad Gita, the Hindu holy book—it did not annihilate the world, of course. The blast produced a crater something like 10 feet deep and 2,400 feet in diameter. Its energy was estimated to be approximately 20 kilotons, equivalent to 20,000 tons of TNT.

Explosions are virtually never measured in terms of TeV, because that unit represents a far smaller quantity of energy. Nothing stops us from making the conversion, however, and we determine that the nuclear explosion at Trinity was equivalent to 5.0×10^{20} TeV. That's 5 followed by 20 zeroes—an enormous figure—something like the average number of stars in a billion galaxies. Thus, even the most rudimentary atomic bomb releases astronomically more energy than any of the particle collisions we've discussed.

Only a few weeks after the Trinity test, bombs were dropped over Hiroshima and Nagasaki, leading to the devastation of those Japanese cities and the end of World War II. The dawn of the atomic age brought increased apprehension about the possibility that scientific miscalculation, combined perhaps with political blunders, would lead to Earth's apocalyptic demise. It didn't help matters that brilliant scientists such as Edward Teller and Herman Kahn would dispassionately discuss the effectiveness of new types of nuclear weaponry in scenarios involving mass casualties.

In an era of trepidation, horror films provided a welcome release valve for anxiety. Fictional threats from aliens that could be warded off through concerted effort were easier to handle emotionally than factual dangers from ourselves that seemed to defy ready solution. The 1958 movie *The Blob*, about a rapidly growing creature from outer space, offers a case in point. Its plot is simple: an alien meteorite delivers a gelatinous cargo that turns out to be a ravenous eater. Each time the blob ingests something, it becomes larger. Soon it is humongous and pining for human "snacks" found in a movie theater and a diner. People flee in terror from the gluttonous menace until the film's hero, played by Steve McQueen, manages to freeze it using fire extinguishers.

In a poll about which astronomical objects the public deemed most bloblike, black holes, the compact relics of massive stars, would surely come in first place. The image of such an astronomical object stealthily entering a theater, absorbing all of the patrons, enlarging itself, and moving on to another venue might not be all that far from the popular stereotype. Several black hole properties bring blobs to mind. If a black hole is near an active star, for instance as a binary system, it can gradually acquire matter from the star, due to its mutual gravitational attraction, and become more massive over time. Nothing is mysterious or unusual about this process except that black holes form particularly steep gravitational wells. Astronomers observe this accumulation of material through images of the radiation emitted as it falls inward toward the black hole.

The physics of black holes derives from Einstein's general theory of relativity. In 1915, with the ink on Einstein's gravitational theory barely dry, German physicist Karl Schwarzschild, while serving on the Russian front in the First World War, discovered an exact solution. He solved Einstein's equations for a static, uniform, nonrotating ball of matter, and mapped out the geometry of the space surrounding it. The Schwarzschild solution, as it is called, represents the gravitational influence of simple, spherical astronomical bodies. It describes precisely how a sphere of matter, such as a star or a planet, dents the geometry of space-time and forces nearby objects to move along curved paths. Whether these objects flee, orbit, or crash depends on whether their speeds exceed the escape velocity required to take off. Those nearby objects without enough escape speed, such as an insufficiently fueled rocket, are doomed for impact.

One curious aspect of the Schwarzschild solution that at first seemed simply a mathematical anomaly, but later became the basis of serious astronomical consideration, is that for dense enough objects there exists a spherical shell, called the event horizon, within which nothing that enters can escape, not even light. That's because the escape velocity within the event horizon is faster than the speed of light. Therefore no object can reach such a speed and flee. In the 1960s, John Wheeler coined the term *black hole* to describe such a perpetually dark, ultracompact object. Black holes represent the ultimate dents in the fabric of space-time—the "bottomless pits" of astronomy.

The idea of inaccessible regions of space raises profound questions about the laws of physics in such enclaves. Are physical principles the same inside and outside a black hole's event horizon? How would we know, if no one could venture inside and come back to tell the tale? Wheeler was puzzled in particular by the question of what would happen to disordered matter entering the point of no return. According to the long-established law of entropy, for any closed system, any natural process must either preserve or increase the total amount of entropy. Entropy is a measure of the amount of disordered energy, or waste, in a physical system. Thus, although natural processes can convert ordered energy into waste (such as a forest fire turning a stately grove into ashes), they can never do the trick of transforming waste energy completely into fuel. Although it is an open question whether the law of entropy applies to the cosmos as a whole, Wheeler was troubled by the idea that we could fling our waste into black holes, it would vanish without a trace, and the total fraction of orderly energy in the universe would increase. Could black holes serve as the cosmetic of cosmology, gobble up signs of aging, and make the universe seem more youthful?

In 1972, Jacob Bekenstein, a student of Wheeler's, proposed a remarkable solution to the question of black hole entropy. According to Bekenstein's notion—which was further developed by Stephen Hawking—any entropy introduced by absorbed matter falling into a black hole would lead to an increase in the area of its event horizon. Therefore, with a modest increase in entropy, the event horizon of a black hole would become slightly bigger. The signs of aging in the universe would thereby manifest themselves through the bloating of black holes.

As Hawking demonstrated, Bekenstein's theory offers startling implications about the ultimate fate of black holes. Although nothing can escape a black hole's event horizon *intact*, Hawking astonished the astrophysics community by theorizing that black holes gradually radiate away their mass. Hawking radiation, as it is called, is a natural consequence of another conjecture by Bekenstein. In addition to defining the entropy of black holes, Bekenstein showed that they also have temperature. Because anything in nature with finite temperature, from lava to stars, tends to glow (either visibly or invisibly), Hawking speculated that black holes radiate, too. To evade the event horizon's barrier, this would be a quantum tunneling process, akin to how alpha particles escape the strong attraction of nuclei. A painstakingly slow trickle of particles, much too protracted to observe, would emerge over the eons. The more massive the black hole, the lower its temperature, and the longer it would take for it to completely evaporate. For a black hole produced by a collapsed star ten times the mass of the Sun, it would take an estimated 10^{70} (*1* followed by *70* zeroes) years for the whole thing to radiate away—far, far longer than the age of the universe. Because of such a prolonged time span, Hawking radiation has yet to be observed.

Less massive black holes would evaporate more quickly. These would need to be produced, however, through a completely different mechanism than stellar collapse. Stars of solar mass, for example, end up as the faint objects called white dwarfs, not black holes. Rather than collapsing further, their internal pressure supports their weight and they simply fade away over time.

Nevertheless, the Schwarzschild solution makes no mention of a minimal black hole mass. Rather, it defines a Schwarzschild radius—the distance from the center to the event horizon—for any given mass, no matter how small. The lighter an object, the tinier its Schwarzschild radius.

A black hole ten times solar mass, for instance, would have a Schwarzschild radius of almost nineteen miles, allowing it to fit comfortably within the state of Rhode Island. If an incredibly powerful force could somehow squeeze Earth so that it was smaller than its Schwarzschild radius, it would be only the size of a marble. A human being shrunk down to less than his or her Schwarzschild radius would be billions of times smaller than an atomic nucleus—clearly below the threshold of direct measurement.

In 2001, Savas Dimopoulos, along with Brown University physicist Greg Landsberg, published an influential paper speculating that microscopic black holes could be found at the LHC. These would have Schwarzschild radii comparable to the Planck length, less than 10^{-33} inches—or one quadrillionth of the size of a nucleus. Basing their work on theories of large extra dimensions, the researchers estimated that the LHC would churn out ten million microscopic black holes each year, similar to the annual rate of Z particles that were produced at the LEP.

Dimopoulos and Landsberg pointed out that any microscopic black holes produced at the LHC could be used as delicate tests of the number of extra dimensions in the universe—potentially verifying the braneworld hypothesis that gravitons leak into a

A simulation of the production and decay of a microscopic black hole in the ATLAS detector.

parallel realm. That's because the mass of these minute compact objects depends on how many dimensions space contains. Because Hawking radiation vanquishes lighter bodies at a faster rate, they would evaporate almost instantly, decaying into potentially detectable particle by-products. Their discovery would thereby present an ideal way to study the process of Hawking radiation, as well as examining dimensionality.

The existence of microscopic black holes is at this point purely hypothetical. Stellar-size black holes are still not fully understood, let alone theorized miniature variations. Dimopoulos and Landsberg emphasized that their calculations involved "semiclassical arguments" set in the nebulous zone between general relativity and certain theories of quantum gravity—particularly string theory and M-theory. "Because of the unknown stringy corrections," they wrote, "our results are approximate estimates."[1]

When a subject is as little known as the application of quantum theory to gravity at the smallest scales in nature, it is hard to say for sure which theoretical predictions will yield tangible results. The brilliance of detectors such as ATLAS and CMS is that they are general purpose. Data they collect will be analyzed by various groups all over the world and matched up against all different kinds of hypotheses. Until then, microscopic black holes remain fascinating to consider but highly speculative.

If microscopic black holes do pop up, they would have virtually no time to interact with their environment, which would just be the evacuated, low-temperature, hermetically sealed collision site. Produced by interacting quarks from two colliding protons, they would immediately decay into other elementary particles. During their brief existence, they would weigh little more than heavy atomic nuclei and would be far enough away from everything else that their gravitational interactions would be negligible. No fireworks, or even a blip on a screen, would announce their appearance. The only way anyone at the LHC would recognize that they came and went would be through meticulous data analysis that could well take many months.

Psychological perceptions of risk don't always match up to actuality. Exotic threats, when matched against familiar hazards, often seem much scarier. People don't spend their time worrying about the national injury rate due to slipping on bathroom floors or falling down basement stairs unless it unfortunately happens to their loved ones or to them. Yet there's something about the bloblike image of black holes that stirs apprehension, even if the chances that such objects, particularly on a microscopic scale, will affect people's lives are about as close to zero as you can imagine.

In 2008, Walter L. Wagner and Luis Sancho filed a lawsuit in Hawaii's U.S. District Court seeking a restraining order that would halt operations of the LHC until safety issues involving potential threats to Earth were fully investigated. The named defendants included the U.S. Department of Energy, Fermilab, CERN, and the National Science Foundation. In a twenty-six-page decision, the judge hearing the case dismissed the suit, stating that the court did not have jurisdiction over the matter.

Trained in nuclear physics, Wagner heads a group called Citizens Against the Large Hadron Collider that he has established to warn against potential doomsday scenarios. One such scenario is the production of microscopic black holes that somehow manage to persist. This could happen, he conjectures, if Hawking radiation proves ineffective or nonexistent. After all, he points out, it has never actually been observed. The enduring mini–black hole would either pass right through Earth, like a neutrino, or be captured by Earth's gravity. Suppose the latter is true. Once embedded in the core of our planet, he speculates, it could engorge itself with more and more material, grow bigger and bigger, and threaten our very existence. As Sancho and Wagner describe in their complaint:

> Eventually, all of earth would fall into such growing micro-black-hole, converting earth into a medium-sized black hole, around which would continue to orbit the moon, satellites, the ISS [International Space Station], etc.[2]

This doomsday scenario is reminiscent of the catastrophe described in David Brin's 1990 novel, *Earth*. In that science fiction epic set in the year 2038, scientists create a miniature black hole that accidentally escapes from its magnetic cage. After plunging into Earth's interior, it is primed to gobble up the whole planet. A chase ensues to find the voracious beast before it is too late.

The anti-LHC group urges us not to wait that long. If there is even the slightest chance of a black hole destroying the world, the group argues, why take the risk? Why not rule out all conceivable hazards *before* the particle roulette begins? It would be a compelling case only if mini–black holes could really grow like blobs from microscopic to Earth-threatening sizes—but no credible scientific theory indicates that they could.

Another concern of Wagner's group is the possibility of the LHC engendering "strangelets" or particle clusters with equal numbers of up, down, and strange quarks. According to the strange matter hypothesis, such combinations would be more stable under certain circumstances than ordinary nuclear matter. Like heat changing a runny egg into a solid glob, the energy of the LHC could catalyze such an amalgamation. Then, according to Sancho and Wagner's complaint:

> Its enhanced stability compared to normal matter would allow it to fuse with normal matter, converting the normal matter into an even larger strangelet. Repeated fusions would result in a runaway fusion reaction, eventually converting all of Earth into a single large "strangelet" of huge size.[3]

Yet another purported global threat is magnetic monopoles. These would be magnets with only north or south poles, not both. Chop a bar magnet in half and you get two smaller magnets, each with north and south poles. No matter what, there would always be two poles per magnet. Monopoles, in contrast, would have just one. Dirac predicted their existence in the 1930s, and they are an important component of certain Grand Unified Theories (GUTs).

Although monopoles have never been seen in nature, some theorists anticipate that they would be extremely massive and possibly turn up in LHC debris.

Sancho and Wagner ponder a scenario in which two massive monopoles, one north and the other south, would be produced at the LHC. Interacting with ordinary matter, theoretically they might hasten certain GUT processes and induce protons to decay. Suppose this caused a chain reaction, causing proton after proton—and atom after atom—to disintegrate into other particles. Eventually, the whole world would be a lifeless orb of inhospitable decay products.

Given the story of the Trinity test, such dire scenarios might lead us to believe that CERN researchers are now taking bets on the fate of Earth. Could LHC workers be wagering each lunchtime on whether black holes, strangelets, monopoles, or another bizarre creation will gobble up the French soil as if it were toast, bore holes through the Swiss mountains as if they were cheese, make haste toward Bologna, recklessly slice it up, and still not be satisfied? Could there be a hidden plot to conceal the true dangers of the world's largest collider?

On the contrary, CERN prides itself on its openness. Secrecy is anathema to its mission. Isidor Rabi, who participated in the Manhattan Project and witnessed the Trinity test, founded CERN as a way for Europeans after the war to rebuild peaceful, civilian science on a cooperative basis. He emphasized that CERN would not have nuclear reactors and that none of its findings would be classified—to preclude the possibility that its research could be used for destructive purposes.

Scientists at CERN sometimes joke about the hullabaloo over mini–black holes. With gallows humor, some jest with a wink and a smile about being at the epicenter of imminent catastrophe—then go right back to coding their software. "[Friends] know that I'm not an evil scientist trying to kill the world," said graduate student researcher Julia Gray.[4]

Theorists realize that quantum uncertainty offers a minuscule chance for an astonishingly diverse range of eventualities. Why spend time worrying about these? Through an extraordinarily unlikely roll of quantum dice, Nima Arkani-Hamed remarked that "the Large Hadron Collider might make dragons that might eat us up."[5]

Despite the lighthearted attitude of many of its researchers, though, the CERN organization itself, for the sake of maintaining a candid and amicable relationship with the international community, takes any public fears very seriously. It doesn't want people to suspect that something sinister is going on in its tunnels and caverns.

Although CERN conducted a comprehensive safety study in 2003 that found no danger from mini–black holes, strangelets, and magnetic monopoles, it agreed to complete a follow-up report that was released in June 2008. The new report concurred with the original findings, offering a number of powerful arguments why microscopic black holes, strangelets, and monopoles, if they truly exist, would pose no threat to Earth.

In the case of miniature black holes, the report showed how conservation principles would preclude them from being stable. Following the maxim that anything not forbidden is allowed, if they are produced by elementary particles, they could also decompose into elementary particles. Thus, aside from the question of whether Hawking's description of black hole radiation is correct, mini–black holes must decay.

Moreover, produced in proton-proton collisions, chances are that microscopic black holes would carry positive charge and thereby be repelled by other positive charges on Earth. Hence, they'd have a hard time approaching atomic nuclei. Even if they could somehow survive and overcome the forces of electrical repulsion, their rate of accreting matter through gravitational attraction would be inordinately slow. In short, any Lilliputian blobs produced at the LHC would not have long for this world. They'd be the goners, not us.

Strangelets, the report's authors point out, would be most likely to be produced during heavy ion collisions rather than proton collisions. In fact, some theorists anticipated their production at the Relativistic Heavy Ion Collider (RHIC), a facility at Brookhaven that opened in 2000. Interestingly, Wagner filed lawsuits against that collider too, unsuccessfully trying to prevent it from going on line. Yet during the RHIC's run, absolutely no strangelets have turned up. The nearby Hamptons beach resort continues to attract the rich and famous to its glistening sand and surf, untainted by strange matter. If strangelets never appeared where they were most expected, why worry about them showing up at a less likely place?

There's good reason to expect that any strangelets produced in collisions would be extremely unstable. They'd break up at temperatures much lower than those generated in ion crashes. The report compares the chances of stable strangelet production under such searing conditions to the "likelihood of producing an icecube in a furnace."

Monopoles were explored at length in the 2003 study, as the authors of the 2008 report point out. If they managed to disintegrate protons—an extremely hypothetical scenario supported only in certain GUTs—they could gobble up barely a fraction of a cubic inch of material before being blasted harmlessly into space by the energy produced in the decays. A tiny hole in an LHC detector would be, at most, the only souvenir of their fleeting existence.

Finally, in arguing against the dangers of black holes, strangelets, monopoles, and other hypothetical high-energy hazards, perhaps the CERN safety team's most compelling argument is that if any apocalyptic scenarios could occur, they would have happened already in cosmic ray events. Cosmic rays are enormously more energetic than what the LHC or any other collider provides. As the report points out:

> Over 3×10^{22} cosmic rays with energies of 10^{17} eV or more, equal to or greater than the LHC energy, have struck the Earth's surface since its formation. This means that Nature has already conducted the equivalent of about a hundred thousand LHC experimental programmes on Earth already—and the planet still exists.[6]

If CERN's reassurances aren't enough, perhaps we can be comforted by the lack of warning signals from the future. According to Russian mathematicians Irina Aref'eva and Igor Volovich, the LHC might have the energy to create traversable wormholes in space-time linking the present with the future. If the LHC represented a danger, perhaps, as in the case of Gregory Benford's novel *Timescape* (1980), scientists from the future would relay messages back in time to warn us. Or maybe, as in John Cramer's novel *Einstein's Bridge* (1997), they would try to change history and prevent the LHC from being completed.

A traversable wormhole is a solution of Einstein's equations of general relativity that connects two different parts of space-time. Like black holes, wormholes are formed when matter distorts the fabric of the universe enough to create a deep gravitational well. However, because of a hypothetical extra ingredient called phantom matter (or exotic matter) with negative mass and negative energy, wormholes respond differently to intruders. While matter dropping into a black hole would be crushed, the phantom matter in a traversable wormhole would prop it open and allow passage through a kind of a space-time "throat" to another cosmic region. The difference would be a bit like attempting passage through a garbage disposal versus through an open pipe.

Researchers have speculated since the late 1980s that certain kinds of traversable wormhole configurations could offer the closed timelike curves (CTCs) that permit backward time travel. CTCs are hypothetical loops in space-time in which the forward direction in time of a certain event connects with its own past, like a dog chasing its own tail. For large enough wormholes, by following such a loop completely around, an intrepid voyager (in a spaceship, for example) could theoretically travel back to any time after the CTC's creation. Smaller wormholes would just allow the passage of particles and information. Still, they might allow people to contact younger versions of themselves.

Aref'eva and Volovich conjecture that the LHC's energetic cauldron could brew up wormholes that allow backward communication. LHC researchers might learn of this first if they receive bizarre messages on their computer screens dated years ahead. Perhaps their e-mail in-boxes would suddenly become clogged with spam from companies yet to exist.

As numerous science fiction tales relate, backward time travel could create paradoxes involving the violation of causality: the law that a cause must precede its own effect. For example, suppose an LHC technician discovers a message from future researchers who have discovered how to send interpretable signals backward in time through wormholes created in the machine. The message warns of the creation of a new type of particle that will start to decimate Earth. After receiving the warning, CERN administrators decide to shut down the LHC. In that eventuality, the original future wouldn't exist. How then could the future researchers have sent the signals? It would be an effect (turning off the machine) with either a cause from the future or no cause at all.

In such paradoxical situations, that's where parallel universes would come in handy. If, in a manner similar to the Many Worlds scenario, each time information, things, or people travel backward in time, the universe bifurcates into several versions, the cause in one strand could precipitate an effect in another without contradiction.

Currently, most high-energy physicists have more pressing concerns about the future than hypothetical global disasters or whether backward-traveling signals are possible. When you are running a machine as complex as the LHC, and planning future upgrades and projects, pragmatic considerations generally overshadow abstract speculations. The LHC detectors have so many delicate components, subject to extreme conditions such as temperatures near absolute zero, it takes an incredible amount of effort to make sure they are working properly.

In between tinkering with current technologies, if a high-energy physicist has time to contemplate the future, he or she might well be thinking about the future of the field itself. How will the LHC results—however they turn out—affect the direction of particle physics? At what level of funding and commitment will the public continue to support one of the most expensive scientific disciplines? Is it reasonable to encourage young students to pursue high-energy careers given such uncertainty? What will particle research look like decades from now?

Conclusion

The Future of High-Energy Physics

The International Linear Collider and Beyond

The future of particle physics is unthinkable without intense international collaboration.

—LEV OKUN (PANEL DISCUSSION ON THE FUTURE OF PARTICLE PHYSICS, 2003)

Where to go from here? After more than seventy-five years of breathtaking progress, the future of high-energy physics is by no means certain. Much depends on what is found at the Large Hadron Collider (LHC).

In the most disappointing case, if no new physics were found at the LHC, the physics community would have to rethink its priorities. Would pressing collider energies even higher to probe greater particle masses be worth the cost? In an era of tight budgets, could governments around the world even be convinced to fork over the colossal sums needed to construct new ultra-powerful machines for a possibly illusive search? If the LHC came up empty, mustering political support for an even larger device would be an unlikely prospect. "It will probably be the end of particle physics,"[1] said Martinus Veltman, referring to the possibility of the Higgs existing but not turning up at the LHC.

There's no reason to expect such a bleak outcome, however. Assuming that the LHC does discover new particles, such as the Higgs or supersymmetric companions, heralds potential sources of dark matter, opens the door to an exploration of new dimensions, and/or finds something altogether unexpected, the theorists would work through the data and determine which models the results support. Then they would assess what new information would be needed to fill in any gaps.

Ideally, novel findings at the LHC would help decide to what extent the Standard Model is an accurate depiction of nature at a wide range of energy levels. It could also rule out the purest version of the Standard Model in favor of supersymmetric theories or other alternatives. Whittling down theories to the likeliest possibilities would be a happy outcome indeed. If past experience is any judge, however, given the creativity of theorists, there could well be more alternatives than ever before. What to do then?

Because of the Superconducting Super Collider (SSC) debacle, the prospect of American laboratories picking up where CERN leaves off are extremely poor. Aside from contributions to European and international projects, which have proven extremely vital, American high-energy physics looks cloudy in general. No new accelerator labs are in the planning, and the existing ones are grappling with severe reductions in funding.

Fermilab has been living on borrowed time for more than two decades. When the decision was made to locate the SSC in Texas, researchers in Batavia braced themselves for the final spins of their beloved particle carousel—even before the ride truly started. Fate would extend the merry-go-round's run, however. The SSC's cancellation in 1993 and the discovery of the top quark at the Tevatron in 1995 showed why the latter was so critical to particle physics. Instead of the machine shutting down permanently, it was temporarily closed for a thorough upgrade.

From 1996 until 2000, a $260 million makeover aspired to transform the Tevatron into an even more stunning collider. After the surgical stitches were taken out, a second series of experiments began, called Run II, and researchers examined the results of the face-lift. Run II produced many notable achievements, including enhanced measurements of the top quark mass, lower bounds on the Higgs mass, and examinations of hadrons containing bottom quarks. Yet, during the early part of the 2000s, its collision rate was not as high as hoped. Fermilab's directorship realized that to maximize the chances for important discoveries before the LHC went online, they would have to get the machine running even more efficiently.

Fortunately, further efforts during scheduled shutdowns in winter 2004–2005 and in spring 2006 boosted the Tevatron's luminosity to record levels. To accomplish this extraordinary feat, machine experts integrated the Recycler antiproton storage ring (a method for accumulating antiprotons) more effectively into the Tevatron workings and used the method of electron cooling to tighten up the antiproton beam. This granted the Tevatron yet a few more years of life.

Once the LHC is fully operational, though, the impetus to preserve the Tevatron will largely disappear. With a maximum energy of 2 TeV, it is unlikely that discoveries would be made with the Tevatron that aren't found first with the LHC. The only way to significantly increase the Tevatron's energy would be to build a new ring, which is not in the cards. Furthermore, the Tevatron's use of antiprotons impedes its luminosity compared to proton-proton colliders such as the LHC. Antiprotons are much harder to engender than protons, given that the latter are easily mass-produced from ordinary hydrogen gas. In short, after midwifing particle events far longer than expected, the Tevatron could well be on the brink of retirement. As postdoctoral researcher Adam Yurkewicz jokingly remarked, the Tevatron has been "running so long, puffs of steam are coming out."[2]

Given the uncertain future of U.S. laboratories, young American researchers planning to enter the field of high-energy physics had best expect to spend much time in Europe—or alternatively to conduct all of their research remotely. Either possibility has potential drawbacks. Traveling back and forth to Europe can be hard on families and also—if unfunded by stipends—on the wallet. To be safe, researchers anticipating spending time in Geneva might wish to choose partners and friends who are international diplomats, bankers, or fondue chefs—wealthy ones, preferably.

The alternative, conducting all of one's research remotely, also has its perils. If researchers-in-training are based at CERN during a time in which hardware is being installed or repaired, they might gain valuable knowledge about instrumentation. But if they are spending their prime educational years at a remote institution that happens to be connected to CERN's computational Grid, they might never have a chance. Suppose a graduate student never has hardware experience and specializes exclusively in computer analysis. He or she becomes a postdoctoral researcher and continues to concentrate in perfecting software packages. Then comes time for a professorship. Would a university be willing to gamble on hiring an experimentalist who doesn't know a thing about calibrating calorimeters or wiring up electronics?

The concentration of the field of high-energy physics in just a handful of labs—and soon perhaps in a single site—coupled with the rise of larger, more complex detectors has effectively reduced the possibilities for direct experience with the tangible aspects of the field. The days of sitting in trailers parked near tunnels waiting for telltale signals—an emblem of experimental work during the latter decades of the twentieth century—have come to a close. Instead, except for those lucky enough to be present during the building or upgrading of detectors, high-energy physics is largely becoming a hands-off occupation. Given that physical measurements are now conducted in supercooled chambers hundreds of feet beneath the ground, where radiation exposure can be perilous, such a progression is logical. Yet, will sitting in front of computer monitors, either in Geneva or elsewhere, and running statistical software be an exciting enough enterprise to attract the next generation of high-energy physicists?

In the mid-2010s, hands-on expertise will once again be key, when the LHC completes a planned upgrade to what is sometimes called the Super Large Hadron Collider. The main purpose of the enhancement is to boost the machine's luminosity and increase the rate of productive collisions even further. When the collider is shut down for the upgrade, the detectors will also be taken apart. Burned-out electronics, baked by years of radiation damage, will be replaced and other instrumentation upgraded to improve the detectors' performance.

Aside from the Super LHC, the next great hope for particle physics is an exciting new project called the International Linear Collider (ILC). As its name suggests, it is the first collider to be planned and funded by the international community, rather than mainly by the United States or Europe. The Superconducting Super Collider (SSC) was *supposed* to be international, but that never quite worked out. CERN accepts funding from beyond the European community specifically to support detector projects (ATLAS, CMS, and so forth) but not for the machines themselves. Thus if the ILC transpires, it would represent a milestone for global scientific endeavors.

The ILC is planned to be twin linear accelerators facing each other—one energizing electrons and the other positrons—housed in a tunnel more than twenty miles long. The reason for its linearity is to avoid energy losses due to synchrotron radiation—a major problem for high-speed orbiting electrons and positrons but not for those traveling in a straight path. To accelerate bunches of these particles close to light-speed, more than eight thousand superconducting niobium radio frequency cavities (perfecting conducting metal sheets used to transfer radio frequency energy to particles), each more than three feet long, will deliver a series of thirty million volt kicks.

All told, these will boost the electrons and positrons up to 250 GeV each. Hence when they collide they will yield 500 GeV of energy, some of which will transform into massive particles. A vertex detector at the collision site will track the decay products of anything interesting that is produced.

Electron-positron collisions are relatively clean and thus ideal for precise measurements of mass. Consequently, though the ILC will be much less energetic than the LHC, its utility will be in pinning down the masses of any particles discovered at the more energetic device. For example, if the LHC produces a potential component of dark matter, the ILC will weigh it and thus inform astronomers what chunk of the cosmos might consist of that ingredient. Knowing the density of the universe would then offer clues as to its ultimate fate. Thus, the ILC would offer a valuable high-precision measuring device—a kind of electronic scale for the world of ultraheavy particles.

So far, the ILC is still in its early planning stages. A site has yet to be chosen—with countries such as Russia making offers. Coordinating the efforts to attract funding and design the project is Barry Barish, the ILC's director, who formerly led the GEM (Gammas, Electrons, and Muons) project (for the aborted SSC) and the Laser Interferometer Gravitational Wave Observatory. After an initially enthusiastic response to the ILC from many different countries, he has been dismayed that some have started to back away from their prior commitments.

In 2007, after initially supporting research and development of the ILC at the level of $60 million, Congress abruptly reduced funding to $15 million. By October of that year, the ILC had spent much of the allocation, even though the funding was supposed to last until 2008. In a December news release, Barish noted, "The consequences for ILC are dire."[3]

Many Europeans are frustrated that American support for scientific projects is unreliable. "In the U.S. everything has to be approved on a year-to-year basis," said physicist Venetios Polychronakos. "Nobody is going to trust the U.S. to be a partner."[4]

After years of reliability, U.K. funding for science has also gone wobbly. In December 2007, Britain's Science and Technology Facilities Council (STFC) issued a report with dreary news for the ILC. "We will cease investment in the International Linear Collider," it stated. "We do not see a practicable path towards the realisation of this facility as currently conceived on a reasonable timescale."[5]

Recalling, perhaps, a happier age when British nuclear scientists were the "champions of the world," Queen guitarist-turned-astronomer Brian May decried the funding cuts. Addressing a ceremonial gathering honoring his appointment as chancellor of Liverpool John Moores University, he said, "I think it is a big mistake and we are putting our future internationally at risk in science. . . . We need support for the great scientific nation we have been."[6]

Because of the U.S. and U.K. decisions, the ILC is by no means a sure thing. Much depends on a restoration of the commitment by wealthier countries to pure science. Given the global economic crisis, funding for basic research has been a tough sell. Perhaps discoveries made by the LHC will attract enough interest to bolster support for a new collider. If the ILC is to avoid the same fate as ISABELLE and the SSC, its proponents will need to make the strongest possible case that precise measurements of massive particles will be critical to the future of physics.

Though there is much speculation, there are no concrete plans in the works for colliders more energetic than the LHC. Conceivably, the CERN machine will prove the end of the line. In the absence of new accelerator data, physicists would lose an important means of testing hypotheses about the realm of fundamental forces and substances. Astronomical measurements of the very early universe, through detailed probes of the microwave background—higher-precision successors to the Wilkinson Microwave Anisotropy Probe survey perhaps—would become the main way of confirming field theories. Perhaps the ultimate secret of unifying all the forces of nature would be found that way.

Until the day when colliders are a thing of the past, let's celebrate the glorious achievements of particle physics and wish the LHC a long and prosperous life. We herald the extraordinary contributions of Rutherford, Lawrence, Wilson, Rubbia, and so many others in revealing the order and beauty of the hidden subatomic kingdom. May the LHC open up new treasure vaults and uncover even more splendor. Like Schliemann's excavations of Troy, the deeper layers it unearths should prove a sparkling find.

Notes

Introduction. The Machinery of Perfection

1. Steven Weinberg, in Graham Farmelo, "Beautiful Equations to Die For," *Daily Telegraph*, February 20, 2002, p. 20.
2. Bryce DeWitt, telephone conversation with author, December 4, 2002.
3. President William J. Clinton, Letter to the House Committee on Appropriations, June 16, 1993.
4. Lyn Evans, "First Beam in the LHC—Accelerating Science," CERN press release, September 10, 2008, press.web.cern.ch/press/PressReleases/Releases2008/PR08.08E.html (accessed March 2, 2009).
5. Peter Higgs, in "In Search of the God Particle," *Independent*, April 8, 2008, www.independent.co.uk/news/science/in-search-of-the-god-particle-805757.html (accessed April 18, 2008).
6. Lyn Evans in "Meet Evans the Atom, Who Will End the World on Wednesday," *Daily Mail*, September 7, 2008, www.mailonsunday.co.uk/sciencetech/article-1053091/Meet-Evans-Atom-end-world-Wednesday.html (accessed March 4, 2009).
7. J. P. Blaizot et al., "Study of Potentially Dangerous Events during Heavy-Ion Collisions at the LHC: Report of the LHC Safety Study Group," *CERN Report 2003-001*, February 28, 2003, p. 10.

1. The Secrets of Creation

1. Isaac Newton, *Opticks*, 4th ed. (London: William Innys, 1730), p. 400.

2. The Quest for a Theory of Everything

1. L. M. Brown et al., "Spontaneous Breaking of Symmetry," in Lillian Hoddeson et al., eds., *The Rise of the Standard Model: Particle Physics in the 1960s and 1970s* (Cambridge: Cambridge University Press, 1997), p. 508.

3. Striking Gold: Rutherford's Scattering Experiments

1. Mark Oliphant, "The Two Ernests, Part I," *Physics Today* (September 1966): 36.
2. David Wilson, *Rutherford, Simple Genius* (Cambridge, MA: MIT Press, 1983), p. 62.
3. J. J. Thomson, *Recollections and Reflections* (New York: Macmillan, 1937), pp. 138–139.
4. Ernest Rutherford to Mary Newton, August 1896, in Wilson, *Rutherford, Simple Genius*, pp. 122–123.
5. Ernest Rutherford to Mary Newton, February 21, 1896, in ibid., p. 68.
6. Thomson, *Recollections and Reflections*, p. 341.
7. Arthur S. Eve, in Lawrence Badash, "The Importance of Being Ernest Rutherford," *Science* 173 (September 3, 1971): 871.
8. Chaim Weizmann, *Trial and Error* (New York: Harper & Bros., 1949), p. 118.
9. Ibid.
10. Ernest Rutherford, "The Development of the Theory of Atomic Structure," in Joseph Needham and Walter Pagel, eds., *Background to Modern Science* (Cambridge: Cambridge University Press, 1938), p. 68.
11. Ibid.
12. Ernest Rutherford to B. Boltwood, December 14, 1910, in L. Badash, *Rutherford and Boltwood* (New Haven, CT: Yale University Press, 1969), p. 235.
13. Ernest Rutherford to Niels Bohr, March 20, 1913, in Niels Bohr, *Collected Works*, vol. 2 (Amsterdam: North Holland, 1972), p. 583.
14. Werner Heisenberg, *Physics and Beyond: Encounters and Conservations* (New York: Harper & Row, 1971), p. 61.

15. Niels Bohr, in Martin Gardner, *The Whys of a Philosophical Scrivener* (New York: Quill, 1983), p. 108.

4. Smashing Successes: The First Accelerators

1. George Gamow, *My World Line: An Informal Autobiography* (New York: Viking, 1970), pp. 77–78.
2. Ibid.
3. Glenn T. Seaborg and Richard Corliss, *Man and Atom: Building a New World through Nuclear Technology* (New York: Dutton, 1971), p. 24.
4. Leo Szilard, "Acceleration of Corpuscles," Patent Application, December 17, 1928, in Valentine Telegdi, "Szilard as Inventor: Accelerators and More," *Physics Today* 53, no. 10 (October 2000): 25.
5. "Man Hurls Bolt of 7,000,000 Volts," *New York Times*, November 29, 1933, p. 14.
6. A. Brasch and F. Lange, *Zeitschrift für Physik* 70 (1931): 10–11, 17–18.
7. Herbert Childs, *An American Genius: The Life of Ernest Orlando Lawrence* (New York: Dutton, 1968), pp. 40–41.
8. Ernest O. Lawrence and J. W. Beams, "The Element of Time in the Photoelectric Effect," *Physical Review* 32, no. 3 (1928): 478–485.
9. Mark L. Oliphant, "The Two Ernests-I," *Physics Today* (September 1966): 38.
10. Childs, *An American Genius*, pp. 146–147.
11. N. P. Davis, *Lawrence and Oppenheimer* (New York: Da Capo, 1986), p. 28.
12. Childs, *An American Genius*, pp. 139–140.
13. E. T. S. Walton, "Recollections of Nuclear Physics in the Early Nineteen Thirties," *Europhysics News* 13, no. 8/9 (August/September 1982): 2.
14. Ibid.
15. E. T. S. Walton to Winifred (Freda) Wilson, April 17, 1932, in Brian Cathcart, *The Fly in the Cathedral: How a Group of Cambridge Scientists Won the International Race to Split the Atom* (New York: Farrar, Straus and Giroux, 2004), p. 238.
16. Oliphant, "The Two Ernests-I," p. 45.
17. Childs, *An American Genius*, p. 210.
18. Oliphant, "The Two Ernests-I," p. 44.

5. A Compelling Quartet: The Four Fundamental Forces

1. Emilio Segrè, *Enrico Fermi, Physicist* (Chicago: University of Chicago Press, 1970), p. 72.
2. Cecil Powell, *Fragments of Autobiography* (Bristol: University of Bristol Press, 1987), p. 19.
3. Gamow, *My World Line*, p. 127 (see chap. 4, n. 1).

6. A Tale of Two Rings: The Tevatron and the Super Proton Synchrotron

1. Ernest Lawrence, reported by Robert R. Wilson, interviewed by Spencer Weart on May 19, 1977, in Robert Rathbun Wilson, "From Frontiersman to Physicist," *Physics in Perspective* 2 (2000): 182.
2. Robert R. Wilson, interviewed by Spencer Weart on May 19, 1977, in ibid., p. 183.
3. Ibid, p. 173.
4. Brookhaven National Laboratory, "The Cosmotron," www.bnl .gov (accessed April 28, 2008).
5. Andrew Sessler and Edmund Wilson, *Engines of Discovery: A Century of Particle Accelerators* (Singapore: World Scientific, 2007), p. 59.
6. Peter Rodgers, "To the LHC and Beyond," *Physics World* 17 (September 1, 2004): 27.
7. Philip J. Hilts, *Scientific Temperaments: Three Lives in Contemporary Science* (New York: Simon & Schuster, 1982), p. 23.
8. J. D. Jackson, "Early Days of Wine and Cheese," *Fermilab Annual Report* (Batavia, IL: Fermi National Accelerator Laboratory, 1992).
9. Robert P. Crease and Charles C. Mann, *The Second Creation: Makers of the Revolution in 20th-Century Physics* (New York: Collier Books, 1986), p. 343.
10. Robert R. Wilson and Adrienne Kolb, "Building Fermilab: A User's Paradise," in Lillian Hoddeson et al., eds., *The Rise of the Standard Model: Particle Physics in the 1960s and 1970s* (Cambridge: Cambridge University Press, 1997), pp. 356–357.
11. M. Bodnarczuk, ed., "Reflections on the Fifteen-Foot Bubble Chamber at Fermilab" (Batavia, IL: Fermilab, 1988).
12. Burton Richter, "The Rise of Colliding Beams," in Lillian Hoddeson et al., *The Rise of the Standard Model*, p. 263.
13. Crease and Mann, *The Second Creation*, p. 345.

14. Dieter Haidt, "The Discovery of the Weak Neutral Currents," *AAPPS Bulletin* 15, no. 1 (February 2005): 49.
15. Gary Taubes, "Carlo Rubbia and the Discovery of the W and the Z," *Physics World* (January 9, 2003): 23.
16. Daniel Denegri, "When CERN Saw the End of the Alphabet," *CERN Courier* 43, no. 4 (May 1, 2003): 30.
17. "Europe 3, U.S. Not Even Z-Zero," *New York Times*, June 6, 1983, p. A16.
18. Boyce D. McDaniel and Albert Silverman, *Robert Rathbun Wilson (1915–2000), Biographical Memoirs*, vol. 80 (Washington, DC: National Academy Press, 2001), p. 12.
19. Herwig Schopper, telex to Leon Lederman, July 5, 2003, Fermilab Archives.

7. Deep in the Heart of Texas: The Rise and Fall of the Superconducting Super Collider

1. Lillian Hoddeson and Adrienne W. Kolb, "The Superconducting Super Collider's Frontier Outpost, 1983–1988," *Minerva* 38 (2000): 275.
2. Leon Lederman with Dick Teresi, *The God Particle: If the Universe Is the Answer, What Is the Question?* (New York: Houghton Mifflin, 2006), p. 379.
3. Attributed to Jack London. See, for example, Irving Shepard, *Jack London's Tales of Adventure* (New York: Doubleday, 1956), p. vii.
4. George F. Will, "The Super Collider," *Washington Post*, February 15, 1987, p. C7.
5. "Remarks by the President in Meeting with Supporters of the Superconducting Super Collider Program," March 30, 1988, Fermilab History Collection.
6. Sharon Begley, "The 'Quark Barrel' Politics of the SSC," *Newsweek*, July 2, 1990.
7. "Yes, Big Science. But Which Projects?" *New York Times* editorial, May 20, 1988, p. A30.
8. Robert Reinhold, "Physics, Shymics—This Project Is a $6 Billion Plum," *New York Times*, March 29, 1987, p. D4.
9. Dietrick E. Thomsen, "States Race SSC Site-Proposal Deadline—Superconducting Super Collider Site Selection," *Science News*, September 12, 1987.

10. Ben A. Franklin, "Texas Is Awarded Giant U.S. Project on Smashing Atom," *New York Times*, November 11, 1988, p. A1.
11. Venetios Polychronakos, in discussion with the author at CERN, June 27, 2008.
12. Leon Lederman, in "Exodus of Scientists Forecast at Fermilab," *New York Times*, November 15, 1988, p. C8.
13. Tom Kirk to Michael Riordan, June 17, 1999, in Lillian Hoddeson and Adrienne W. Kolb, "The Superconducting Super Collider's Frontier Outpost, 1983–1988," *Minerva* 38 (2000): 304.
14. Michael Riordan, "A Tale of Two Cultures: Building the Superconducting Super Collider," *Historical Studies in the Physical and Biological Sciences* 32 (2001): 129.
15. Neil Ashcroft, in William J. Broad, "Heavy Costs of Major Projects Pose a Threat to Basic Science," *New York Times*, May 27, 1990, p. A1.
16. Arno Penzias, in Malcolm Browne, "Supercollider's Rising Cost Provokes Opposition," *New York Times*, May 29, 1990, p. A1.
17. Michael Riordan, "The Demise of the Superconducting Super Collider," *Physics in Perspective* 2 (2000): 416.
18. Raphael Kasper, in Sherry Jacobson, "Superconductor Staff Reunites 10 Years Later," *Dallas Morning News* (Ellis County), July 23, 2005, p. 1.
19. William John Womersley, in Charles Seife, "Physics Tries to Leave the Tunnel," *Science* 302, no. 5642 (October 3, 2003): 36.
20. Wade Roush, "Colliding Forces: Life after the SSC," *Science* 266, no. 5185 (October 28, 1994): 532.
21. Eric Berger, "Behind a Scientific Success, a Failed Texas Experiment," *Houston Chronicle*, May 25, 2008, p. 1.
22. Jeffrey Mervis, "Scientists Are Long Gone, But Bitter Memories Remain," *Science* 302, no. 5642 (October 3, 2003): 40.
23. Lederman, *The God Particle*, p. x.

8. Crashing by Design: Building the Large Hadron Collider

1. CERN Communication Group, "LHC the Guide," CERN-Brochure-2008-001-Eng, p. 31.

9. Denizens of the Dark: Resolving the Mysteries of Dark Matter and Dark Energy

1. Vera Rubin, in Amy Boggs, "AAVC Award for Distinguished Achievement," *Vassar, The Alumnae/i Quarterly* 103, no. 1 (Winter 2006).

2. Vera Rubin, in Stephen Cheung, "Alum Vera Rubin receives award for achievements in astrophysics," *Miscellany News (Vassar College)*, February 15, 2007.

10. The Brane Drain: Looking for Portals to Higher Dimensions

1. Voltaire, *Candide*, trans. Shane Weller (Mineola, NY: Dover, 1991), p. 5.
2. Bryce DeWitt, telephone conversation with author, December 4, 2002.
3. C. B. Collins and S. W. Hawking, "Why Is the Universe Isotropic?" *Astrophysical Journal* 180 (March 1, 1973): 171.

11. Microscopic Black Holes: A Boon to Science or a Boom for the World?

1. S. Dimopoulos and G. Landsberg, "Black holes at the LHC," *Physical Review Letters* 87 (2001): 161602.
2. "Complaint for temporary restraining order, preliminary injunction, and permanent injunction," Luis Sancho and Walter L. Wagner, plaintiffs, Civil No. 08-00136-HG-KSC, U.S. District Court of Hawaii.
3. Ibid.
4. Julia Gray, conversation with author at CERN, June 26, 2008.
5. Nima Arkani-Hamed, in Dennis Overbye, "Asking a Judge to Save the World, and Maybe a Whole Lot More," *New York Times*, March 29, 2008, p. D1.
6. "Review of the Safety of LHC Collisions," *LHC Safety Assessment Group*, CERN Report, June 2008.

Conclusion. The Future of High-Energy Physics: The International Linear Collider and Beyond

1. Martinus Veltman, in J. R. Minkel, "As LHC Draws Nigh, Nobelists Outline Dreams—and Nightmares," *Scientific American*, July 2, 2008, www.sciam.com/article.cfm?id=as-lhc-draws-nigh-nobelis (accessed July 13, 2008).
2. Adam Yurkewicz, conversation with author at CERN, June 26, 2008.
3. Barry Barish, "Director's Corner," *ILC Newsline*, December 20, 2007, www.linearcollider.org/cms/?pid=1000466 (accessed July 13, 2008).

4. Venetios Polychronakos, conversation with author at CERN, June 27, 2008.
5. Science and Technology Facilities Council, "Delivery Plan," December 11, 2007, www.scitech.ac.uk/resources/pdf/delplan_07.pdf (accessed July 14, 2008).
6. Brian May, in Roger Highfield, "Brian May Attacks Science Research Cuts," *Daily Telegraph*, April 14, 2008, www.telegraph.co.uk/earth/main.jhtml?view=DETAILS&grid=&xml=/earth/2008/04/14/scibrian114.xml (accessed July 13, 2008).

Further Reading

Technical works are marked with an asterisk.

Asimov, Isaac. *Atom: Journey Across the Subatomic Cosmos* (New York: Dutton, 1991).

Boorse, H. A. *The Atomic Scientists: A Biographical History* (New York: John Wiley & Sons, 1989).

Brown, A. *The Neutron and the Bomb: A Biography of James Chadwick* (New York: Oxford University Press, 1997).

Brown, Laurie, and Lillian Hoddeson, eds. *The Birth of Particle Physics* (Cambridge: Cambridge University Press, 1983).

Cathcart, Brian. *The Fly in the Cathedral: How a Group of Cambridge Scientists Won the International Race to Split the Atom* (New York: Farrar, Straus and Giroux, 2004).

Chamberlain, Basil Hall, trans. *The Kojiki: Records of Ancient Matters* (Rutland, VT: C. E. Tuttle, 1982).

Childs, Herbert. *An American Genius: The Life of Ernest Orlando Lawrence* (New York: Dutton, 1968).

Cockburn, S., and D. Ellyard. *Oliphant. The Life and Times of Sir Mark Oliphant* (Adelaide, Australia: Axiom Books, 1981).

Crease, Robert P., and Charles C. Mann. *The Second Creation: Makers of the Revolution in 20th-Century Physics* (New York: Collier Books, 1986).

Davies, Paul. *Superforce: The Search for a Grand Unified Theory of Nature* (New York: Simon & Schuster, 1984).

Fermi, Laura. *Atoms in the Family: My Life with Enrico Fermi* (Chicago: University of Chicago Press, 1954).

Gamow, George. *My World Line: An Informal Autobiography* (New York: Viking, 1970).

Greene, Brian. *The Elegant Universe: Superstrings, Hidden Dimensions, and the Quest for the Ultimate Theory* (New York: Vintage Books, 2000).

———. *Fabric of the Cosmos: Space, Time and the Texture of Reality* (New York: Knopf, 2004).

Guth, Alan. *The Inflationary Universe: The Quest for a New Theory of Cosmic Origins* (Reading, MA: Perseus, 1998).

Halpern, Paul. *Cosmic Wormholes: The Search for Interstellar Shortcuts* (New York: Dutton, 1992).

———. *The Great Beyond: Higher Dimensions, Parallel Universes and the Extraordinary Search for a Theory of Everything* (Hoboken, NJ: John Wiley & Sons, 2004).

Halpern, Paul, and Paul Wesson. *Brave New Universe: Illuminating the Darkest Secrets of the Cosmos* (Washington, DC: National Academies Press, 2006).

Hilts, Philip J. *Scientific Temperaments: Three Lives in Contemporary Science* (New York: Simon & Schuster, 1982).

Hoddeson, Lillian, Laurie Brown, Max Dresden, and Michael Riordan, eds. *The Rise of the Standard Model: Particle Physics in the 1960s and 1970s* (Cambridge: Cambridge University Press, 1997).

Hoddeson, Lillian, Adrienne Kolb, and Catherine Westfall. *Fermilab: Physics, the Frontier, and Megascience* (Chicago: University of Chicago Press, 2008).

Jaspers, Karl. *The Great Philosophers*, vol. 3, trans. Edith Ehrlich and Leonard H. Ehrlich (New York: Harcourt Brace, 1993).

Kaku, Michio. *Parallel Worlds: A Journey through Creation, Higher Dimensions and the Future of the Cosmos* (New York: Doubleday, 2004).

Kane, Gordon. *The Particle Garden: Our Universe as Understood by Particle Physicists* (Reading, MA: Helix Books, 1995).

King, Leonard W., trans. *The Seven Tablets of Creation, or the Babylonian and Assyrian Legends Concerning the Creation of the World and of Mankind* (London: Luzac, 1902).

Lederman, Leon, with Dick Teresi. *The God Particle: If the Universe Is the Answer, What Is the Question?* (New York: Houghton Mifflin, 2006).

*Lee, S. Y. *Accelerator Physics* (Singapore: World Scientific, 2004).

Oliphant, Mark. *Rutherford—Recollections of the Cambridge Days* (London: Elsevier, 1972).

Pais, Abraham. *Inward Bound: Of Matter and Forces in the Physical World* (New York: Oxford University Press, 1986).

Patterson, Elizabeth C. *John Dalton and the Atomic Theory* (New York: Doubleday, 1970).

Randall, Lisa. *Warped Passages: Unraveling the Mysteries of the Universe's Hidden Dimensions* (New York: HarperCollins, 2005).

*Randall, Lisa, and Raman Sundrum. "An Alternative to Compactification," *Physical Review Letters* 83 (1999): 4690–4693.

Rhodes, Richard. *The Making of the Atomic Bomb* (New York: Touchstone, 1988).

Riordan, Michael. *The Hunting of the Quark: A True Story of Modern Physics* (New York: Simon & Schuster, 1987).

Schumm, Bruce A. *Deep Down Things: The Breathtaking Beauty of Particle Physics* (Baltimore: Johns Hopkins University Press, 2004).

Segrè, Emilio. *Enrico Fermi, Physicist* (Chicago: University of Chicago Press, 1970).

Sessler, Andrew, and Edmund Wilson. *Engines of Discovery: A Century of Particle Accelerators* (Singapore: World Scientific, 2007).

Smith, P. D. *Doomsday Men: The Real Dr. Strangelove and the Dream of the Superweapon* (New York: St. Martin's Press, 2007).

Snow, C. P. *The Physicists: A Generation That Changed the World* (London: Papermac, 1982).

'T Hooft, Gerard. *In Search of the Ultimate Building Blocks* (New York: Cambridge University Press, 1997).

Thomson, J. J. *Recollections and Reflections* (New York: Arno Press, 1975).

Weinberg, Steven. *Dreams of a Final Theory: The Scientist's Search for the Ultimate Laws of Nature* (New York: Vintage, 1992).

———. *The First Three Minutes: A Modern View of the Origin of the Universe* (New York: Basic Books, 1993).

Wilson, D. *Rutherford: Simple Genius* (London: Hodder, 1983).

*Wilson, Edmund. *An Introduction to Particle Accelerators* (New York: Oxford University Press, 2001).

Wouk, Herman. *A Hole in Texas* (New York: Little, Brown, 2004).

Index

Page numbers in *italics* refer to illustrations.

accelerators
- Cockcroft and Walton's research and, 93–95
- Cockcroft-Walton generator and, 80–82, *81*
- Cockcroft's attempt to break nuclear targets and, 77
- complex realm of particles on, 115
- definition of, 13–14
- early experiments with, 75–98
- at Fermilab, 128, 131
- Gamow's quantum tunneling formula and, 78
- Ising's prototype of, 79–80
- Lawrence's cyclotron and, 90–92, *91,* 93, 95, 108, 117, 118
- lightning strikes as, 85–86
- Powell's construction of, 106–7
- recycling of, 164, 167
- Superconducting Super Collider (SSC) and, 156–57
- Van de Graaff's work with, 83–85, 90
- Wideröe's design of, 128–29
- Wideröe's ray transformer and, 78–79, 86, 89

ADD model, 202–4, 205
Adelberger, Eric, 204
ADM formulation, 114
air (element), 24
Akeley, Lewis, 87
ALICE (A Large Ion Collider Experiment) detector, 5, 19, 168, 174, 175–76
alpha particles, 58–60, 62–66, 68, 72, 78
Alpher, Ralph, 109
Alternate Gradient Synchrotron (AGS), 121
Ampère, André-Marie, 31
ancient Greece, 24, 29
Anderson, Carl, 93, 105
Andromeda galaxy, 181–82
angular momentum, 67
anthropic principle, 196–97

antigravity, 189
antimatter, 13, 15, 42, 73
antineutrinos, 102
antiprotons, 137, 142, 227
antiquarks, 14, 137, 142
Antoniadas, Ignatius, 202
Aref'eva, Irina, 222, 223
argon, in particle detectors, 169
Aristotle, 24
Arkani-Hamed, Nima, 202, 220
Arnowitt, Richard, 114
Ashcroft, Neil, 159
astronomy, 178, 180
ATLAS (A Toroidal LHC ApparatuS) detector, 1–5, 18–19, 168, 169–70, 171, *171*, 172, 174, 176–77, 206, *215*, 216, 229
atomic bomb, development of, 100
atomic number, 65–66
atomism, 24–25, 28–30
atoms
  ancient Greek concepts of, 24, 30
  Dalton's work with, 25, 58
  electrostatic force and, 29
  first use of term, in modern sense, 25
  Lawrence's time interval measurements involving, 88, 89
  nucleus of, 65–66
  radioactive processes and, 58
  relative weights of, 25
  solar system comparison with, 27
  Thomson's "plum pudding" model of, 61, 62, 65
attractive forces, 28
axions, 186
Bardeen, John, 44
Barish, Barry, 158, 169, 230
baryons, *103*, 137
Beams, Jesse, 88
Becquerel, Henri, 58
Bednorz, Johannes, 158
Bekenstein, Jacob, 213–14
Benford, Gregory, 222
Berkeley Radiation Laboratory ("Rad Lab"), 92, 118–19, 187
beta decay, 93, 101–103, 112, 113
beta particles, 58, 61
Bethe, Hans, 108, 109
Bevatron, 120
Big Bang conditions
  dark energy scenarios and, 188
  general theory of relativity on, 37–38
  measurement of background radiation left over from, 38
  particle detectors to reproduce conditions of, 13, 19, 22, 39, 42
Big Bang theory, 38, 109, 110–11
Big Crunch, 188
Big Rip, 189
Big Whimper, 188
blackbodies, 35
Blackett, Patrick, 73, 96
black holes
Bekenstein's theory on expansion of, 213–14
  first use of term, 213
  Large Hadron Collider (LHC) research and creation of, 20–21, 22
  MACHOs (Massive Compact Halo Objects) and, 184
  microscopic, 216, 220–21
  physics of, 212
  public concern over, 212

Blewett, John, 120
Bohr, Niels, 66–68, 72, 96, 98, 99–100, 115, 121
Born, Max, 69
bosons, *103*
  asymmetry involving fermions and, 12
  beginning of the universe and, 42
  as category of elementary particles, 44
  exchange particles and, 102
  intermediate vector bosons, 131
  Standard Model prediction of, 121
  string theory and, 49
  supersymmetry for uniting fermions and, 139–40
  Yukawa's electromagnetic research and, 104–5
bottom quarks, 14, 19, 130, 136, 141
Boyle, Robert, 24–25, 29
Boyle's Law, 25
braneworld hypothesis, 192–93, 199
Brasch, Arno, 85
Breit, Gregory, 90
bremstrahlung, 127
Brin, David, 218
Brobeck, William, 120
Brout, Robert, 45
brown dwarfs, 183–84
bubble chambers, Fermilab, 125–26, *126*, 128, 133
Burstein, David, 183
Bush, George Herbert Walker, 153, 155

Calabi, Eugenio, 201
Calabi-Yau spaces, 201
Caldwell, Robert, 189
calibration, 144
calorimeters, 18, 126, 127–28, 143, 144, 169, 170–71, 172
carbon, 108, 109
Carter, Brandon, 196
cathode rays, 57
Cavendish, Henry, 54–55
Cavendish Laboratory, Cambridge
  Chadwick's research on neutrons at, 73–74, 92–93, 97
  Cockcroft and Walton's splitting of a lithium nucleus at, 93–95
  Cockcroft-Walton generator at, 74, 80–82, *81*
  cyclotron proposal for, 97
  description of, 54–55
  Gamow's research at, 75, 76–77
  Rutherford as director of, 55–56, 72, 74, 82, 96, 97, 98, 106
  Thomson as director of, 55, 56, 66, 72
  Walton's linear accelerator at, 80–82
CDF (Collider Detector at Fermilab) Collaboration, 143–44, 144–45
Central Design Group (CDG), 150
Central Tracking Chamber, Tevatron, 143
CERN (European Organization for Nuclear Research)
  description of, 1, 5–6
  founding of, 219
  funding of, 17
  hardware knowledge of researchers at, 228

CERN (*Continued*)
location of, 1, 5, 22, 121
public fears of work of, 220–22
Chadwick, James, 73–75, 92–93, 96, 97
Charge-Parity (CP) violation, 175
charginos, 186
charm quarks, 14, 130
Cherenkov, Pavel, 127
Cherenkov detectors, 126, 127, 144
Chu, Paul, 158
Citizens Against Large Hadron Collider, 217
Cline, David, 132, 134
Clinton, Bill, 16, 160
closed strings, 202
closed timeline curves (CTCs), 222
cloud chambers, 73, 106, 143
CMS (Composer Muon Solenoid) detector, 5, 19, 168, 172, *173*, 174, 206, 216, 229
COBE (Cosmic Background Explorer) satellite, 110, 195
Cockcroft, John Douglas, 77, 80–82, 91, 93–95, 96
cold dark matter, 186
Collider Detector at Fermilab (CDF) Collaboration, 143–44, 144–45
Coma Cluster, 180
Collins, C. B., 195–97
Congress, and Superconducting Super Collider (SSC), 16–17, 152, 153, 154, 156, 159, 160
conservation of parity, 113
Cooper, Leon, 44
Cooper pairs, 44
corpuscles (small particles), 24–25, 28–29, 57, 61. *See also* electrons
Cosmic Background Explorer satellite, 187
cosmic rays, 221
cosmological constant, 37–38, 189
Cosmotron, 120–21
Coulomb, Charles-Augustin de, 29, 31
Coulomb's law, 31
Courant, Ernest, 120, 121
Cowan, Clyde, 112
CP (Charge-Parity) violation, 175
Cronin, James, 175
cryostats, 170
Curie, Marie, 58, 83, 96
Curie, Pierre, 58

D0 Collaboration, Tevatron, 144
Dai, J., 201
Dalton, John, 25–26, 58
dark energy
definition of, 180, 189
gravitational theories on, 190–91
interest in mystery of, 178, 186–87
Large Hadron Collider (LHC) and, 179–80, 186, 190
Supernova Cosmology Project (SCP) on, 187–88
dark matter, 179–86
axions and, 186
cold dark matter and, 186
definition of, 179–80
gravitational theories on, 190–91
hot dark matter and, 185–86
interest in mystery of, 52, 178, 180–81
Large Hadron Collider (LHC) and, 190
MACHOs (Massive Compact Halo Objects) and, 183–85, 186

WIMPs (Weakly Interacting Massive Particles) and, 185, 186
Dave, Rahul, 189
de Broglie, Louis, 70, 96
deceleration of the universe, 188
decupole magnets, 167
Democritus, 24, 30
Denegri, Daniel, 138
Department of Energy (DOE), 149–50, 151, 152, 153, 154, 155, 159, 217
Deser, Stanley, 114
deuterons, 95
deuterium, 60–61, 92, 108
DeWitt, Bryce, 13, 194, 197
Dicke, Robert, 110
Dimopoulos, Savas, 50–51, 202, 215, 216
dipole magnets, 167
Dirac, Paul, 72, 96, 199, 218
Dirichlet brane (D-brane), 201–2
Displacement Law, 61
Doppler shifts, 182, 187
down quarks, 14, 130
drift chambers, 126, 128
dualities, 200
duality of dualities, 200–201
Duff, Michael, 199–200
Duffield, Priscilla, 124
Dvali, Gia, 202
Dyson, Freeman, 111

earth (element), 24
Edlefsen, Niels, 91–92
Einstein, Albert
cosmological constant theory of, 37–38, 189
on electron "spin-down," 71
equivalence of energy and mass equation of, 10, 14, 82, 95
equivalence principle and, 36–37
general theory of relativity of, 11–12, 37, 190, 194, 212
Heisenberg's matrix mechanics and, 68
photoelectric effect and, 36, 66, 88
quantum mechanics and, 72
Rutherford compared to, 62
special theory of relativity of, 34, 36, 46
speed of light research by, 34
Szilard's letter to Roosevelt signed by, 100
unification model sought by, 42–43, 49, 99–100
electromagnetic calorimeters, 4, 127, 143, 170–71, 172
electromagnetic interaction, 29, 111
electromagnetic radiation, 32–34
electromagnetic waves, 31, 32
electromagnetism, 8–9, 12, 30–32, 42–43, 47, 54, 104–5
electron-positron colliders, 129–30, 230
electron synchrotrons, 119–20
electrostatic force, 29–30, 31
electroweak unification. See Standard Model
element-building models, 185
elements
ancient Greek concepts of, 24
atomist views of, 24–25
Boyle's use of term, 24
Dalton's work with, 25, 58
Mendeleyev's listing of, in table form, 26

elements (*Continued*)
predicting discovery of new elements, 26
theories on production of, 109
Empedocles, 24, 190
"Energy Production in Stars" (Bethe), 108
Engler, François, 45
equivalence principle, 36–37
ether hypothesis, 33–34
Eureka (Poe), 8
European Organization for Nuclear Research. *See* CERN
Evans, Lyn, 17–18, 19, 21
Eve, Arthur, 61
Everett, Hugh, 193–94
Everett, Kenneth, 122–23
exchange particles, 9, 102, 137
exclusion principle (Pauli), 44, 71

false vacuum, 10
Faraday, Michael, 30, 78
Fermi, Enrico, 100–3, 112, 113, 210
Fermi National Accelerator Laboratory (Fermilab), 123
accelerators at, 128, 131, 134
detector design at, 169
establishment of, 108, 122
future of, 226–27
HPWF collaboration to find the W boson at, 132
Wilson's design of, 122–25
Wilson's directorship at, 134, 141
fermions, 102, *103*
asymmetry involving bosons and, 12
beginning of the universe and, 42
as category of elementary particles, 44
neural current events involving, 133
string theory and, 49
superpartners to, 140
supersymmetry for uniting bosons and, 139–40
Fermi weak coupling constant, 102–3
Feynman, Richard, 50, 111–12, 115, 193
Feynman diagrams, 112, 113
fire (element), 24, 32
Fitch, Val, 175
fixed-target accelerators, 128–20
Ford, Kent, 181–83
Franklin, Benjamin, 29
Friedman, Alexander, 38
Frisch, Otto, 100
Fry, Jack, 133–34

galactic recession, law of, 187
galactic rotation curve, 182
gamma decay, 74
gamma radiation, 33, 58
Gammas, Electrons, and Muons (GEM) group, 158, 230
Gamow, George, 75–77, 78, 96, 109, 181
Geiger, Hans, 63–65, 74
Geiger counter, 63
GEM (Gammas, Electrons, and Muons) group, 158
General Dynamics, 157
general theory of relativity, 11–12, 37, 190, 194, 212
Georgi, Howard, 50–51
Glashow, Sheldon, 9, 47, 50, 131
gluinos, 140, 186
gluons, 14, *103*, 140
"God particle." *See* Higgs boson

gold-foil experiments in radiation, 52, 64–65
Goldhaber, Gerson, 187
Goldhaber, Maurice, 187–88
Goudsmit, Samuel, 71
Grand Unified Theories (GUTs), 139, 149, 218–19, 221
gravitons, *103,* 199, 202, 215
gravitational microlensing, 184
gravity (gravitation)
  ADD model and, 202–3
  deceleration of the universe and, 188
  Einstein's equivalence principle on, 36–37
  galaxies held together by, 180
  hierarchy problem linking other forces to, 114
  microscopic black holes and, 216
  natural interactions involving, 9, 12
  Newton's research on, 28, 29, 36
  quantum theory applied to, 114–15
  string theory on, 49
  supersymmetry research on differences between other interactions and, 13, 49
Gray, Julia, 219
Greece, ancient, 24, 29
Green, G. Kenneth, 120
Green, Michael, 50
Grid global computing network, 4–5, 19, 228
Guralnik, Gerald, 45
Guth, Alan, 197–98

hadron colliders, 130
hadronic calorimeter, 4, 127–28, 143
hadrons, 14, *103,* 105, 127–28, 133, 164, 172
Hafstad, Lawrence, 90
Hagen, C. Richard, 45
Hahn, Otto, 100
Haidt, Dieter, 133–34
Hawking, Stephen, 195–97, 213–14, 220
Hawking radiation, 21, 216
heavy hydrogen, 60
Heisenberg, Werner, 68–69, 70–71, 72, 96, 115
helium, 60, 66, 108, 109
hermeticity, 136
Hernandez, Paul, 125
Herschel, William, 32
Hertz, Heinrich, 33, 54
hierarchy problem, 114, 202
Higgs, Peter, 10, 21, 45–47
Higgs boson, *103*
  CERN particle detector research on, 15, 19, 48
  description of, 10–11
  Higgs's work with, 3, 10–11, 45–47
  Large Hadron Collider (LHC) search for, 48, 162, 173–74, 178, 226
  lepton collider in search for, 141
  nickname of "God particle" for, 11, 46
  original reception to first publication of research by Higgs on, 46–47
  possibility of multiple Higgs particles, 19–20, 48
  Standard Model prediction of, 131–32, 139, 140, 149, 226
Higgs field. *See* Higgs boson
Higgs mechanism, 45–47
Higgs particle. *See* Higgs boson

high-temperature superconductivity, 158
Hoddeson, Lillian, 148
hot dark matter, 185–86
Hoyle, Fred, 109, 110
Hubble, Edwin, 38, 109, 180, 182, 187
Hughes, James, 199
hydrogen, 66, 108, 126, 179, 227

IBM, 158
induction, 30, 78
inflation, 198
infrared radiation, 32
inner detector, 169
Interacting Storage Rings (ISR), 130
intermediate vector bosons, 131
International Linear Collider (ILC), 229–31
inverse-square laws, 27
ionization, for tracking charged particles, 143
ISABELLE collider, 139, 148, 169
Ising, Gustav, 79–80, 82, 83
isotopes, 60–61, 115

Jackson, J. David, 124
Joliot-Curie, Irene and Frédéric, 96
J/psi particle, 130, 187

Kahn, Herman, 211
Kaluza, Theodor, 206
Kaluza-Klein excitation, 206, 207
kaons, *103*, 113
Kasper, Raphael, 160
Kepler, Johannes, 182
Kerst, Donald, 129
Kibble, Tom, 45
Kirschner, Robert, 186
Klein, Oskar, 50, 201, 206
Kolb, Adrienne W., 148

Landsberg, Greg, 215, 216
Lange, Fritz, 85
Large Electron Positron Collider (LEP), 16, 140–41, 149–50, 164, 167
large extra dimension, 201, 203
Large Hadron Collider (LHC), 163–78
- American researchers at, 16
- antiprotons and, 227
- black hole research and, 215–16
- braneworld hypothesis and, 192–93
- CERN and, 163–64
- completion of, 15, 17
- damage from helium leak in, 17–18
- dark energy and, 190
- dark matter and, 179–80, 186, 190
- decision to build, 164–65
- description of, 2, 18
- detectors in, 168–69, 174–75
- electricity in region drained by, 166–67
- funding of, 17, 152, 165
- future uses of high-end physics and findings of, 225–26, 232
- Higgs particle research using, 19–20, 162, 173–74, 178
- law suits to halt operations of, 21–22, 217–19
- limitations of, 177
- location of, 15–16, 22, 165–66
- magnets in, 17, 18, 44, 164, 167, 168, 170, 171, *171*, 176–77
- research on origins of the universe and conditions in, 19, 22, 39, 42

string theory and research at, 52
tunnel design in, 167–69
undercurrent of fear about potentially dangerous events from operations at, 20–22
Large Magellanic Cloud, 184
Lattes, César, 107
law of galactic recession, 187
law of multiple proportions, 25
Lawrence, Ernest Orlando, 86–92, 95–98, 232
atomic process time interval measurements of, 88, 89
cyclotron of, 90–92, *91,* 93, 95, 108, 117, 118
family background of, 86–87
University of California, Berkeley, research of, 88–89, 92, 97–98, 119, 120
Lawrence Berkeley National Laboratory, 92, 150, 186, 187
lead ions, in particle detectors, 19
Lederman, Leon, 136, 141, 142, 148, 149, 151, 154, 162
Lee, Tsung Dao, 113–14, 175
Leibniz, Gottfried, 191–92, 193
Leigh, R. G., 201
Lemaitre, Georges, 109
length contraction, 34
lepton colliders, 141
leptons, 14, 42, 102, *103,* 105, 128, 130, 144, 202
Leucippus, 24
LHC. *See* Large Hadron Collider
LHCb (Large Hadron Collider beauty) particle detector, 5, 19, 169, 174
LHCf (Large Hadron Collider forward) detector, 174, 176
light
Einstein's research on, 34, 35, 36
electric charges and, 32
invisible region of, 32
rainbow of colors of, 32
speed of, 34, 35, 36
wavelength measurement of, 32
lightning strikes, as accelerators, 85–86
Linde, Andrei, 198
lithium, 80, 93, 95, 109
Liu, Jun, 199
Livingston, M. (Milton) Stanley, 93, 120
lodestone, 29
Lofgren, Edward, 120
London, Jack, 151

MACHO Project, 184
MACHOs (Massive Compact Halo Objects), 183–85, 186
magnetic fields, 30, 31, 57
magnetic monopoles, 2, 218–19
magnetism, 29
magnets
at Fermilab, 124–25
hadrons as, 164
at Large Hadron Collider (LHC), 17, 18, 44, 164, 167, 168, 170, 171, *171,* 176–77
in Superconducting Super Collider (SSC), 150, 156, 157
in synchrotrons, 119–20, 120–21
Manhattan Project, 100, 119, 124, 210–11, 219
Mann, Alfred K., 132
Manyfold Universe, 203

Many World Interpretation, 194–95
Marsden, Ernest, 63–65, 72, 74, 77
matrix mechanics, 68, 69, 70
Maxwell, James Clerk, 8, 30–32, 33, 34, 54, 55, 115
May, Brian, 231
McGill University, 59–60, 61–62, 100
McIntyre, Peter, 134, 150`
Meitner, Lise, 100
membrane theory, 200
Mendeleyev, Dmitri, 26
mesons, *103,* 105, 106, 108, 113, 127, 130, 137
Michelson, Albert, 34
microlensing, gravitational, 184
Milky Way, 180, 182, 184
Minimal Supersymmetric Standard Model (MSSM), 50–51, 203
Minkowski, Hermann, 36
Misner, Charles, 114
missing matter mystery, 52
Mitchell, Maria, 181
momentum, in atomic research, 69
Morley, Edward, 34
M-theory, 13, 200–201
Müller, Karl, 158
multiple proportions, law of, 25
muon neutrinos, 130
muons, *103,* 105–6, 107, 112, 130
muon system, 4, 171

Nambu, Yoichiro, 44–45
National Academy of Engineering, 153
National Academy of Sciences, 153
National Science Foundation, 217
neutralinos, 186
neutrinos, *103,* 172, 185–86
  detection of, 4, 74, 206
  Fermi's beta decay theory and, 101, 112
neural current events involving, 133
neutrons, *103*
  Chadwick's research on, 73–74, 92–93
  Fermi's beta decay theory and, 101, 102, 112
  Lawrence's estimate of mass of, 96
  particle detectors and, 4, 115, 127
  similar in mass to protons, 43
neutron stars, 184
"New Genesis" (Gamow), 109–10
Newton, Sir Isaac, 27–30, 69, 98
New York Times, 84, 121, 139, 152–53
Nobel Prizes, 58, 63, 72, 108, 131, 187
nuclear fission, 99, 100
nucleosynthesis, 185

Occhialini, Giuseppe, 107, 108
octupole magnets, 167
Oddone, Pier, 141
Oliphant, Mark, 97, 98
O'Neill, Gerald, 129–30
"On Physical Lines of Force" (Maxwell), 31
Oort, Jan, 180, 181
Oort Limit, 180
open strings, 201–2
Oppenheimer, J. Robert, 210–11
Ørsted, Hans Christian, 30
oxygen, 109

pair production, 127
parallel universe idea, 192
parity, 113, 114, 175

particle detectors
  calibration of, 144
  description of, 3–4
  Grid global computing network in, 4–5, 19, 228
  high-energy physics and, 126–28
  particles detected in, 4
  particles not detected in, 4
Pauli, Wolfgang, 70–71, 74, 96, 101
Pauli exclusion principle, 44, 71
Penzias, Arno, 110, 159
Peoples, John, 141
periodic table, 26, 65–66, 115
Perl, Martin, 130
Perlmutter, Saul, 186, 187, 188
phase angle, 10
phase transition, 44, 45
photinos, 186
photoelectric effect, 36, 66, 88
photomultipliers, 126, 127
photons
  early research on, 8–9
  as an example of a boson, 44, 47, *103*
  particle detectors and, 4
  quantum electrodynamics (QED) on, 111, 113
  in string theory, 202
  superpartners of, 140
  Yukawa's electromagnetic research and, 104–5
pions, *103*, 108
Planck, Max, 36, 66
Planck length, 48
Planck scale, 51, 203
Planck's constant, 36, 66, 67, 88
Planck time, 51
plutonium bomb, 210
Poe, Edgar Allen, 8
Polchinski, Joseph, 199, 201, 202
polonium, 58
Polychronakos, Venetios, 230
position, in atomic research, 69
positrons, 72–73, 93, 112, 140
potentials, 45
Powell, Cecil, 106–7, 108
Price, Larry, 1–2
Priestley, Joseph, 29

quadrupole magnets, 167
quanta, 36
quantum chromodynamics (QCD), 137, 139, 186
quantum electrodynamics (QED), 111–13, 114
quantum gravity, 51, 114
quantum mechanics, 11, 26, 44, 51, 69, 72, 194
quantum theory of electromagnetic interaction, 111, 131
quantum theory of matter and energy, 36, 44
quantum tunneling formula, 78
quark-gluon plasma, 175
quarks, 14, 42, 102, *103*, 128, 130, 137, 140, 142, 202, 216
quenching, 177
quintessence, 189–90

Rabi, Isidor I., 106, 121, 219
radioactive decay, 58
radioactivity
  atomic particle research using, 58
  Becquerel's experiments on uranium and, 58
  Marie Curie's research on, 58
  Rutherford's research on alpha particle and, 58–59
radio waves, 33
radium, 58, 60, 63

"Rad Lab" (Berkeley Radiation Laboratory), 92, 118–19, 187
Ramond, Pierre, 49
Ramsay, William, 60, 63
Randall, Lisa, 204–5
Randall-Sundrum model, 204–5
ray transformer, 78–79, 86, 89
Reagan, Ronald, 150–52
Recycler antiproton storage ring, 227
red dwarfs, 184
Reference Designs Study (RDS), 149–50
Reines, Frederick, 112
Relativistic Heavy Ion Collider (RHIC), 220–21
religion, and atomist views, 29
reverse beta decay, 112
Richter, Burton, 130
Riess, Adam, 186
Riordan, Michael, 155
Ritter, Johann, 32–33
Rochester, George, 108
Roentgen, Wilhelm, 33, 58
Roosevelt, Franklin D., 100
rotation curves of galaxies, 182, 183
Rutherford, Ernest, *54,* 232
  accelerator at Cavendish and, 74, 80, 82, 83, 94
  alpha particles research of, 58–60, 62–66, 68, 72, 115
  on atomic power, 82
  as Cavendish Laboratory director, 55–56, 72, 74, 82, 94, 96, 97, 98, 106
  cyclotron proposal and, 97
  death of, 98
  Einstein compared to, 62
  family background of, 53–54, 55–56
  Gamow's research and, 76–77, 78, 80
  lithium nucleus splitting and, 94
  at McGill University, 59–60, 61–62, 100
  proton research and, 72–73
Rubbia, Carlo, 131, 132, 134, 135, 136, 137, 164, 174, 232
Rubin, Vera Cooper, 181–83
Run II (D0 experiment), 227

Salam, Abdus, 9, 47, 131
Sancho, Luis, 217–19
satellites, 110, 187, 195–96
Sceptical Chemist, The (Boyle), 24
Schmidt, Brian, 186
Schopper, Herwig, 142
Schrieffer, J. Robert, 44
Schrödinger, Erwin, 70–71, 72, 96
Schrödinger's position, 70
Schwarz, John, 50
Schwarzschild, Karl, 212
Schwarzschild radius, 214–15
Schwarzschild solution, 212–13, 214
Schwinger, Julian, 111, 115
Schwitters, Roy, 155, 159
Science and Technology Facilities Council (STFC), 231
scintillation counters, 126, 127
SDC (Solenoidal Detector Collaboration), 157
Segrè, Emilio, 101
selectrons, 140
Seventh Solway Conference (Brussels, 1933), 95–96, 101
sextupole magnets, 167
Shane, Donald, 90
signatures, 5

Silicon Vertex Tracker, Tevatron, 143
SLAC, 130, 169
sleptons, 186
Smoot, George, 187
Soddy, Frederick, 60–61
Solenoidal Detector Collaboration (SDC), 157
space-time, 36–37
spark chambers, 126, 128
SPEAR ring, 130
special theory of relativity, 34, 36, 46
spectral lines, 60, 66, 67
"spin-up" and "spin-down," 71
squarks, 140, 186
Stabler, Ken, 152
stacking, 130–31
standard candles, 187
Standard Model
  CERN teams researching, 132–34
  description of electroweak unification and, 9–10, 11, 42–43
  development of, 47
  Higgs particle research and possible revisions to, 19–20, 47–48, 139, 226
  Minimal Supersymmetric Standard Model (MSSM) and, 50–51
  possible bosons predicted by, 121–22, 139, 140
  spontaneous symmetry breaking concept and, 43–44
standing wave, 70
Stanford Linear Accelerator Center, 149, 187
stars
  energy production in, 108
  missing matter dilemma and, 180
state of inertia, 27
Steinhardt, Paul, 189
Stern, Otto, 91
strangelets, 20, 218, 220–21
strange quarks, 14, 130
Strassmann, Fritz, 100
string theory
  bosons in, 49
  braneworld hypothesis and, 199–202, 215
  particle detector research and, 17
  supersymmetry and, 13, 49–50, 52
  as a unification model, 48–50, 52
strong focusing, 121
strong force, 104
strong interactions
  natural interactions involving, 9, 12
  supersymmetry research on differences between other interactions and, 13
Sulak, Larry, 132–33
Sundrum, Raman, 204–5
superconducting magnets, 43–44, 150
Superconducting Super Collider (SSC), 16–17, 147–62
  abandoned site of, 147–48, 161–62
  accelerators and synchrotrons in, 156–57
  cancellation of, 160–61, 164, 226
  federal support for, 150–52, 155
  funding of, 155–56, 160, 229
  location of, 153–54
  magnets in, 150, 156, 157
  opposition to, 158–59
  planning and design of, 149–50
  rationale for building, 149

superconductivity, 42–43
supergroups of researchers, 135–36
Super Large Hadron Collider, 229
Supernova Cosmology Project (SCP), 187–88
Supernova Ia, 187
supernovas, 187
superpartners, 139–40
Super Proton Synchrotron (SPS), 134, 135, 136, 137, 138, 140, 148, 164
superstring theory, 50
supersymmetry (SUSY)
 gravity and, 49
 Higgs particle research and, 12, 13, 178
 implications for Standard Model of, 19–20, 49
 Minimal Supersymmetric Standard Model (MSSM) and, 50–51
 string theory and, 13, 49–50, 52
Swann, W. F. G., 87–88
symmetry
 origins of the universe and clues in, 42–43
 Standard Model and concept of breaking of, 43–44
synchrotrons
 early examples of, 119–21
 fixed-target, 128–20
 in Superconducting Super Collider (SSC), 157
Szilard, Leo, 82–83, 100

tau leptons, 130, 136
tau neutrino, 130, 139
tauons, *103*
Teller, Edward, 211
Tesla coils, 90
Tevatron, Fermi National Accelerator Laboratory, 14–15, 17, 118, 134–35, 141–44, 148, 149, 154, 205, 226, 227
Texas Accelerator Center, 150
Texas National Research Laboratory Commission (TNRLC), 153, 154
Thomson, J. J. (Joseph John)
 Bohr and, 66
 at Cavendish Laboratory, 55, 56, 66, 72
 "plum pudding" model of atoms of, 61, 62, 65
 research on elementary particles by, 57–58, 61
Thomson, Thomas, 25
Thonnard, Norbert, 183
't Hooft, Gerard, 132
thorium, 60
Tigner, Maury, 149, 150, 154, 155
time, as the fourth dimension, 36
time dilation, 34
time-reversal invariance, 113
Tomonaga, Sin-Itiro, 111, 115
top quarks, 14–15, 130, 139, 140, 141, 144–45, 226
TOTEM (TOTal Elastic and diffractive cross-section Measurement), 174, 176
Townshend, Paul, 199–200
transformers, 78–79
Trilling, George, 157
Trinity test, Manhattan Project, 210–11, 219
tritium, 61
true vacuum, 10
Turner, Michael, 189
Tuve, Merle, 86, 87, 90, 91, 96

UA1 (Underground Area 1), 136, 137, 138, 142
UA2 (Underground Area 2), 136, 137, 138, 142
Uhlenbeck, George, 71, 108
ultraviolet catastrophe, 35
ultraviolet radiation, 33
uncertainty principle, 69
U.S. Department of Energy (DOE), 149–50, 151, 152, 153, 154, 155, 159, 217
Universal Fermi Interaction, 112
Universities Research Association (URA), 154–55
up quarks, 14, 130
upsilon, 141
uranium, 58, 60, 99, 100
Urban, Kurt, 85
Urey, Harold, 92

Van de Graaff, Robert Jemison, 83–85, 90, 91
Van de Graaff generator, 84, 85, 90
Van der Meer, Simon, 130–31, 135, 174
Van Hove, Leon, 135
velocities of galaxies, 182, 183
Veltman, Martinus, 132, 226
Villard, Paul, 33, 58
void (element), 24
Volovich, Igor, 222, 223
Voltaire, 191–92

W^+ and W^- bosons, 47, 103, 131, 132, 133, 134, 137–38, 140, 141
Wagner, Walter L., 217–19, 221
Walton, Ernest Thomas Sinton, 78, 80–82, 91, 93–95, 96
water (element), 24
Watkins, James, 159
wave function collapse, 71
wavelengths, color, 32
wave mechanics, 70
wave phenomena, 33, 182
weak exchange particles, 47
weak interaction
  Fermi's beta decay theory and, 101, 113
  Higgs particle research on, 11
  natural interactions involving, 9
  supersymmetry research on differences between other interactions and, 13
  symmetries in, 113–14
  unification of electromagnetism and, 9–10, 11, 42–43
Weinberg, Steven, 9, 13, 47, 131, 132
Weizmann, Chaim, 62
Westinghouse, 157
Wheeler, John, 99, 193, 194–95, 213
Wideröe, Rolf, 78–80, 82, 83, 86, 89, 90, 128–29
Wilkinson Microwave Anistropy Project, 190, 231
Willis, William, 158, 169
Wilson, Charles, 73, 106
Wilson, Robert R. (Bob), 118–25, 232
  at Berkeley Radiation Laboratory ("Rad Lab"), 118–19
  at Cornell University Laboratory for Nuclear Studies, 119, 120, 121–22
  cyclotron design and, 119
  family background of, 118
  Fermilab design of, 122–25
  Fermilab directorship of, 134, 141
  Tevatron commitment by, 134–35, 141

Wilson, Robert W., 110
WIMPs (Weakly Interacting Massive Particles), 185, 186
winos, 140
Witherell, Michael, 141
Witten, Ed, 50, 200–201
WMAP (Wilkinson Microwave Anisotropy Probe) satellite, 110, 195–96, 198
Womersley, William John, 161
wormholes, 222

X rays, 33, 58

Yang, Chen Ning (Frank), 113–14, 175
Yau, Shing-Tung, 201
Young, Charlie, 1–2
Yukawa, Hideki, 103–6, 108, 114

Z boson, 47, 103, *131*, 132, 137, 138, 140, 141
zinos, 140
Zwicky, Fritz, 180, 181

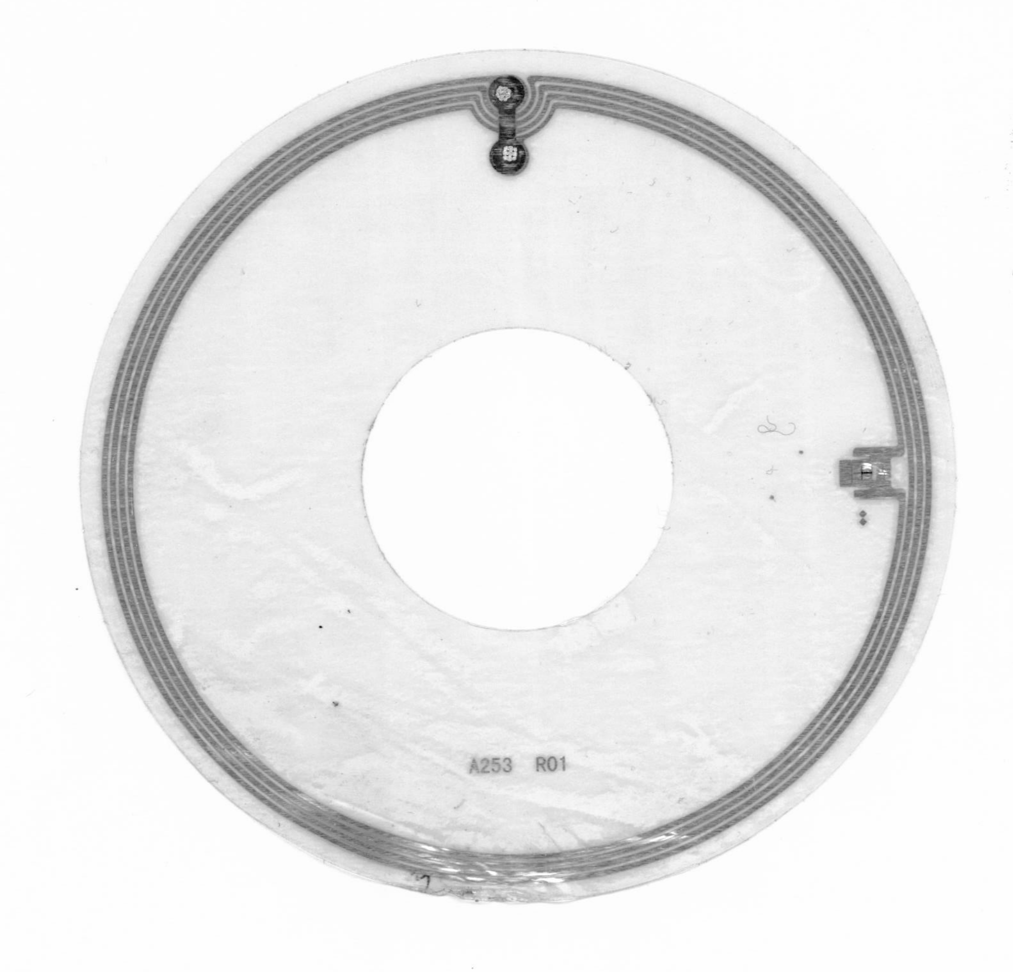
A253 R01